NANOFIBER RESEARCH - REACHING NEW HEIGHTS

Edited by **Mohammed Muzibur Rahman**
and **Abdullah M. Asiri**

Nanofiber Research - Reaching New Heights

http://dx.doi.org/10.5772/61952
Edited by Mohammed Muzibur Rahman and Abdullah M. Asiri

Contributors

Chang-Seop Lee, Yura Hyun, Jeong Hyun Yeum, Graciela Morales, Daniel Grande, Heriberto Rodríguez-Tobías, Tayebeh Behzad, Mobina Ahmadi, Atsuhiro Fujimori, Manami Iizuka, Rie Yamato, YingJie Liao, Takeshi Fukuda, Suxiao Wang, Mitsunori Yada, Lucie Bacakova, Marketa Bacakova, Julia Pajorova, Radmila Kudlackova, Lubica Stankova, Jana Musilkova, Stepan Potocky, Alexander Kromka, Dorota Kolbuk, Gamze Tetik, Fulya Yilmaz, Gizem Celep, Mohammed Muzibur Rahman

CBS, Edition **2018**

Published by InTech

Janeza Trdine 9, 51000 Rijeka, Croatia

Publishing Process Manager Andrea Koric

Technical Editor SPi Global

Cover Designer InTech Design Team

Additional hard copies can be obtained from orders@intechopen.com

Nanofiber Research - Reaching New Heights, Edited by Mohammed Muzibur Rahman and Abdullah M. Asiri
p. cm.
Print ISBN 978-953-51-2528-0
Online ISBN 978-953-51-2529-7

INTECH

open science | open minds

Contents

Preface

It gives us immense pleasure to introduce a book titled "Nanofiber Research - Reaching New Heights" based on state-of-the-art of biomaterials, state-of-the-art of nanocomposites, state-of-the-art of polymers as well as state-of-the-art of nanomaterials and with their various potential and significant applications. These deal with nanotechnological aspects of the synthesis, growth, development and potential applications of nanofibers of materials and polymers. The discussion of these aspects develops through the fundamentals and applied experimental routes in conventional methods through the interaction of conducting materials to finally include the interfacing of the science and technology world. Generally, fiber material has certainly attained achievement of its classical essence and has taken new directions of growth, development, continuous changes and various potential applications in research and development sectors. The new paths and emerging frontiers are branching out from time-to-time around this advanced nanotechnology stage on nanofibers. Advances in fiber materials with instrumentation for evaluating the nanostructure model materials now enable us to understand quite comprehensively almost all the events that take place at the fibers at least at nano level. This work aims to bridge the gap between undergraduate, graduate and researches in applied material as well as polymer sciences, to initiate researchers into nanofiber study in a straightforward way as possible and as to introduce the scientist to the opportunities offered by the applied material science and technology fields.

We worked consistently to accomplish this work on "Nanofiber Research - Reaching New Heights" under the INTECH open access publisher. We hope that this contribution would further enhance the applied fiber materials in polymers and materials science, especially in bringing new entrants into the applied nanofiber science and technology fields and help scientists to forward and develop their own field of specialization.

Dr. Mohammed Muzibur Rahman
CEAMR and Chemistry department,
Faculty of Science, King Abdualziz University,
Jeddah, Saudi Arabia

Prof. Abdullah M. Asiri, (PhD)
Head of Chemistry department,
CEAMR and Chemistry department,
Faculty of Science, King Abdualziz University,
Jeddah, Saudi Arabia

Section 1

Introduction

Chapter 1

Introductory Chapter: Overview of Nanofibers

Mohammed Muzibur Rahman

Additional information is available at the end of the chapter

http://dx.doi.org/10.5772/64370

The book Nanofiber Research - Reaching New heights contains ten chapters divided in four sections. Section 2, "State-of-the-Art Biomaterials," deals with the preparation and characterization of bio-fibers as well as nanofibers for scaffolds as promising cell carrier for tissue engineering. In Section 3, "State-of-the-Art Polymers," the advanced research of electrospinning and electrospraying methods for designing antimicrobial polymeric biocomposite mats and natural medicinal plant polymer for cosmeceutical applications is presented here. The fourth section "State-of-the-Art Nanocomposites" is describing the synthesis and characteristics of carbon nanofibers (CNFs)/silicon composites and application to materials of Li secondary batteries. The final, fifth section "State-of-the-Art Nanomaterials" deals with the preparation, characterization, and potential applications of electrospun metal oxide nanofibers, titanium thin-film nanofibers, and hierarchical crystalline nanofibers on the Cooperative Phenomena of Functional Molecular Group as the Target of Expression of New Physical Properties.

Nanofibers are defined as fibers with diameters less than 100 nm. In the fabric industry, this classification is often extended to include fibers as large as 1,000 nm diameter, which referred to microfibers. Generally, fibers can be prepared by melt processing, interfacial polymerization, electrospinning, antisolvent-induced polymer precipitation, and electrostatic spinning [1], whereas carbon nanofibers are graphitized fibers produced by catalytic synthesis [2]. Here, nanoscale fibers are an interesting noble type of fiber materials that significantly utilized as different doping for momentous applications such as energy storage, fuel cells, aerospace, diodes, capacitors, transistors, drug delivery systems, battery separators, sensors, and information technology [3,4]. Metal oxide nanofibers were made from different sorts of inorganic precursors by different conventional methods. The most regularly mentioned ceramic semiconductor materials with nanofiber surface morphology are TiO_2, SiO_2, lithium titanate, ZrO_2, Al_2O_3, titanium nitride, or platinum. Preparation techniques include direct drawing from a precursor solution, melt processing in furnace, antisolvent-induced polymer precipitation, spinning, electrospinning, or "landmass in the ocean."

One-dimensional magnetic nanostructures have lately attracted much attention due to their intriguing characteristics that are not showed by their bulk or nanoparticle counterparts. These metal oxide nanostructures are also significantly useful as active chemical components for ultrahigh-density data storage in the fabrication of selective and sensitive sensors and spintronic chips or devices [5]. Electrospinning denotes a suitable technique to producing polymer materials along low diameters ranging from nano- to micrometers. The spinning route comprises the application of higher voltage to a micro-capillary that is attached to a reservoir clutching the precursors of polymeric solution. Under the effect of high electric field, a hanging droplet of the monomer solution at the capillary strip can be distorted into the narrowed shape. When the ground strength has outstripped a threshold to overwhelm the surface pressure, electrostatic forces are caused by the discharge of a thin, charged jet. The jet undergoes a whipping technique, which guided to the development of ultrathin fibers as a consequence of solvent evaporation. These nanofibers with significantly moderated diameters foldaway toward a grounded conductor under the accomplishment of the modified electric field and lastly can be accumulated in the form of long threads. Generally, the nanofibers that are deposited on the surface of a conductor form nonwoven mats that are illustrated by higher surface coverage and relatively small pore sizes. By adjusting the design of an accumulator, it is promising to accumulate the thin nanofibers as uniaxially aligned nano-arrays [6–8].

In this book, M. Lizuka et al. focused on "Control of Hierarchical Structure of Crystalline Nanofibers Based on the Cooperative Phenomena of Functional Molecular Group as the Target of Expression of New Physical Properties" where the functional nanofibers were introduced. Here, the formation of nanofiber morphology at a mesoscopic scale and molecular level stacking of a tetrathiafulvalene (TTF) derivative with a chiral group were investigated by the one-dimensional growth method in interfacial molecular films. The growth of morphology is encouraged by the application of the one-dimensional growth method at low-surface pressure conditions, varying compression speeds and sub-phase temperatures. Later on, the formation of nanofiber morphology at a mesoscopic scale and molecular level packing of an amphiphilic diamide derivative with two hydrocarbons were evaluated by the interfacial molecular films. Authors H. Rodríguez-Tobías et al. have discussed about "Electrospinning and Electrospraying Techniques for Designing Antimicrobial Polymeric Biocomposite Mats." Here, they approached on electro-hydrodynamic techniques, namely, electrospraying and electrospinning, which are powerful methods for developing material with morphological features suitable for tissue engineering applications. The incorporation of nanoparticles allows for the generation of multifunctional mats that were used in different applications ranging from filters for pollutant removal to wound dressings. Regarding the pollutant removal, zinc oxide (ZnO) and TiO_2 were the more promising nanoparticles for tissue engineering. Authors J. H. Yeum et al. have focused on "Novel Natural Polymer/Medicinal Plant Extract Electrospun Nanofiber for Cosmeceutical Application" where they approached on zein produced from maize is a hydrophobic protein, which holds great potential for a number of industrial applications, for example, packaging, food, pharmaceutical, cosmetic, and biomedical industries. Here, they studied the influence of cosolvent ratio and concentration of zein/medicinal plant extract on the morphologies of nanostructured zein/medicinal plant extract nanomaterials prepared by electrospinning technique from ethanol/water solution. Zein/medicinal extract nanofibers were characterized in details by field emission scanning electron microscopy, transmission electron microscopy, thermogravimetric analysis, and differential scanning calorimetric techniques. They also worked to incorporate medicinal plant resources into the electrospun zein nanofibers by electrospinning technique to investigate the effect of medicinal extract on the morphologies, antibacterial, antioxidant, and other properties. Authors Y. Liao et al. approached on "Electrospun Metal Oxide Nanofibers and Their Energy Applications" where they focused on metal oxide, a nanofiber which has attracted considerable research interest for processing both one-dimensional nanometer scale morphology and unique chemical and electrical properties. A variety of their practical applications in light-emitting diodes, liquid crystal displays, solar cells, and gas sensors have been demonstrated in this contribution. With the electrospinning technique, it provides a rapid and facile way to fabricate nanofibers with diameter several orders of magnitude smaller than that produced by conventional spinning methods. They also discussed on the fabrication of ultrathin metal oxide nanofibers by the electrospinning technique, where the priority was given to zinc oxide nanofibers. Major parameters affecting the morphology and diameter of the nanofibers were also investigated systematically. The effect of calcination condition on chemical composition and crystallization of the electrospun nanofibers was also addressed in this book. Authors C. S. Lee et al. have focused on "Synthesis and Characteristics of Carbon Nanofibers/Silicon Composites and Application to Materials of Li Secondary Batteries," where they approached on various synthesizing technologies of carbon nanofibers (CNFs) as well as chemical vapor deposition (CVD) technology. It used for hydrocarbon gas or carbon monoxide as a carbon source gas and pyrolyze it to grow CNFs on transition metal catalysts such as Ni, Fe and Co, has been regarded as the most inexpensive and convenient method to produce CNFs for industrial use. Authors D. Kolbuk et al. have described on "Tailoring of Architecture and Intrinsic Structure of Electrospun Nanofibers by Process Parameters for Tissue Engineering Applications." They approached on electrospinning process, which is commercially used to form nanofibers as scaffolds in tissue engineering. Similarities in morphology of electrospun nanofibers to the natural extracellular matrix, flexibility, and low cost of the process contribute to its use in regeneration of the cartilage, ligaments/tendons, muscles, and bones. Properties are tailored by the use of appropriate polymers: polyesters, their copolymers, and blends with natural biopolymers like gelatin, collagen, chitosan, or composites with nanoparticles. In the case of one-component fibers, factors strongly influencing the final diameter of the electrospinning jet include volumetric charge density, distance between the needle and the collector, needle diameter, and viscosity. Authors T. Behzad et al. have focused on "Bio Nanofibers" where they approached on cellulose nanofiber preparation methods from different sources by chemical treatments. They also described the effect of source variations and mechanical processes used in extraction procedure on nanofiber morphology and crystallinity, and chemical composition was evaluated in this contribution. Authors Lucie Bacakova et al. discussed on. "Nanofibrous Scaffolds as Promising Cell Carriers for Tissue Engineering" where they approached on tissue engineering and organs in the human organism, where experimentally used for reconstruction and regeneration of the blood vessels, myocardium and heart valves, skeletal muscle, skin, tendon and ligament, intestine, trachea and bronchi, bladder and urethra, liver, pancreatic islets, brain, spinal cord, optic system, and peripheral nerves. Furthermore, they studied the interaction of human bone-derived cells with nanofibrous scaffolds made of polylactide and its copolymer with polyglycolide, loaded with hydroxyapatite or diamond nanoparticles. They also constructed the novel nanofibers based on diamond deposition on a SiO_2 template and practically demonstrated their effects on the adhesion, viability, and growth of human vascular endothelial cells. Authors M. Yada et al. have focused on "Syntheses and Applications of Titanium Compound Nanofiber Thin Films" where they introduced the generation of titanium compounds with various nanostructures, including titanium phosphate nanobelts and titanate nanofibers on the surface of titanium metal plate, from the treatment of a titanium metal plate in aqueous solutions of various compositions. Various compositions and concentrations and different reaction temperatures and times are combined for chemical processes involving titanium metal and novel titanium compounds/titanium metal composites with various compositions and nanostructures. Here, it was applied research on developing composites that feature the characteristics of both nanostructured titanium compounds and titanium metal. Authors F. Yilmaz et al. have discussed on "Nanofibers in Cosmetics." They developed the cosmetic nanofibers based on the properties on higher surface area to volume ratios, lower diameter, high strength values, low basis weights, higher porosities, and small pore sizes for numerous applications. Excellent interaction with environment, increased loading capacity for agents, high-liquid absorption capacities, high oxygen, and water vapor permeability values are provided by the characteristic properties of nanofibers.

Besides, nanofibers have potential applications in medicine and drug delivery, including artificial organ components, tissue engineering, implant materials [9], injured area dressing, and medical textile nanofiber materials. Lately, scientists have executed that nano-size fiber meshes could be utilized to function with HIV-1 viruses and be capable to be organized as a contraception. In injured area, healing fibers accumulate at the wounded-site and put, depiction the body's particular growth features to the injured-spotted area. Shielding nano-materials including sound absorption materials, shielding clothing over chemical as well as biological warfare agents, and chemi- and bio-sensor applications for identifying toxic and carcinogenic as well as hazardous environmental un-friendly chemical agents were utilized. Nanofibers have also been utilized in pigments for cosmetics as coloring fiber materials. Applications in the textile industry include sport apparel, shoes, climbing, rainwear, outer-wear garments, and baby diapers [10]. Nanofibers comprise antibodies in napkins with several biohazards and chemicals which signal by altering color. Filtration method for potential applications includes heating ventilation and air conditioning (HVAC) system filters, high-efficiency particulate air (HEPA), ultralow penetration air (ULPA) filters, air, oil, fuel filters for automotive, filters for beverage, pharmacy, medical applications, and filter media for new air and liquid filtration applications, such as vacuum cleaners. Fibrous carbon nanomaterials have attracted the substantial attention of researchers [11]. In the last 1960s, carbon nanofibers were introduced as one of the significant industrial implemental nanomaterials for existing science and technology which have been made from different carbon reactant precursors via spinning method. Polyacrylonitrile (PAN) has been utilized as the key reactant precursor with various functionalization as well as modification in producing or processing fibrous materials. Isotropic and anisotropic mesophase pitches and phenolic resins have also been reactant precursors for active carbon nanofibers. A chemical vapor deposition (CVD) method has also formulated carbon nanofibers with a nanostructure and significant chemical and physical properties that are changed from those made via catalytic melt spinning (vapor-grown carbon fiber; VGCF). In the center of VGCFs, thin tubes containing conventional carbon multilayers were executed [12] which were lately reported to be formed by arc discharging [13–15]. The method utilized for the fabrication of VGCFs was effectively functionalized to prepare CNTs [16]. The diameters of CNTs are very low in the nanometer range for single-wall CNTs, in contrast to the μm-range diameters of carbon nanofibers. CNTs are ruminated to be an significant nanomaterial for the improvement of up-to-date nanomaterials, due to carbon nanofibers have substances the improvement of contemporary nanotechnology. Electrospinning has been utilized to make nanofibers of different fiber polymers and diameters. It is a moderately simple method to produce unremitting fibers from fibrous polymer precursor solutions. Carbon fibers are made by electrospinning and carbonization with encapsulation on their nanostructure and characteristic properties in relation to their potential applications, after a short clarification on the circumstance and situations for electrospinning, and some fibrous polymers [17,18].

Nanotechnology of fibrous material is the understanding and controlling of matter at dimensions in nanoscale, where unique phenomena enable novel and significant applications. Nanofibers from cellulose and lignocelluloses play a key role in the nanotechnology field. The development of cellulose nanofibers has attracted significant interest in the last few decades owing to the unique characteristics they endow such as high-active surface-area-to-volume ratio, high Young's modulus, high tensile strength, and low coefficient of thermal expansion [19]. For environmental awareness and the international demand for green technology, bionanocomposites have the potential to replace present petrochemical-based fiber materials. The use of natural fibers instead of traditional reinforcement materials such as glass fibers, carbon, and talc provides several advantages including low density, low cost, good specific mechanical properties, reduced tool wear, and biodegradability [20]. Nanofibers of composites, polymers, and materials are recently used for various significant applications with various compositions in different physical and chemical properties. Nanofibers of doped metal oxides (CdO-ZnO) were prepared by wet chemical method at low temperature in basic medium. Later on the CdO-ZnO, nanofibers were used as photocatalyst for mineralizing the colored reactive Remazol brilliant orange 3R dyes, where it exhibited better photocatalysis than commercially available P25-TiO_2 materials [21]. In vitro, carbon nanofiber microbiosensor were practically developed for selective and sensitive dopamine neurotransmitter and supported copper-graphene oxide nanocomposite onto glassy carbon electrode using conducting coating binders by in situ chemical reduction [22].

Finally, we believe that this book offers broad examples of existing developments in recent development of nanofiber technology research and an excellent introduction to polymer, materials, and composite nanofibers for their potential applications. This work aims to bridge the gap between undergraduate, graduate, and researches in polymers, materials, composite, as well as biomedical sciences, in order to initiate researchers into various nanofiber studies in as straightforward way as possible and to introduce the scientist to the opportunities offered by the applied science and technology fields. It will provide review of the state-of-the-art techniques and advances of fiber researches and bio-nanotechnologies. Perspective authors were invited to present their novel ideas and recent advanced developments in the field of polymers, materials, or related fields. The primary target audience for the book Nanofiber Research - Reaching New Heights includes students, researchers, technologists, physicists, chemists, biologists, engineers, and professionals who are interested in nanofibers and associated topics.

Author details

Mohammed Muzibur Rahman

Address all correspondence to: mmrahmanh@gmail.com

Center of Excellence for Advanced Materials Research (CEAMR) & Chemistry Department, Faculty of Science, King Abdulaziz University, Jeddah, Saudi Arabia

References

[1] Camillo DD, Fasano V, Ruggieri F, Santucci S, Lozzi L, Camposeo A, Pisignano D. Nanoscale, 2013, 5:11637–11642.

[2] Zhou FL, Gong RH. Polymer International. 2008;57:837–845.

[3] Textile World. Nano Technology and Nonwoven. November 2003;P52

[4] Bhat G, Lee Y. *Recent advancements in electrospun nanofibers*. Proceedings of the Twelfth International Symposium of Processing and Fabrication of Advanced Materials, Ed. TS Srivatsan & RA Vain, TMS. 2003.

[5] Li D, Herricks T, Xia Y. Applied Physics Letters. 2003;83:4586.

[6] Theron A, Zussman E, Yarin AL. Nanotechnology. 2001;12:384.

[7] Li D, Wang Y, Xia Y. Nano Letters. 2003;3:1167.

[8] Reneker DH, Chun I. Nanotechnology. 1996;7:216.

[9] Nagy ZK, Zsombor K, Balogh A, Vajna B, Farkas A, Patyi G, Kramarics A, Marosi G. Journal of Pharmaceutical Sciences. 2011;101:322–332.

[10] Pourdeyhimi B. Topic of the month: commercialization potential of nanofiber textile membranes. May 5, 2012.

[11] Inagaki M, Yang Y, Kang F. Advanced Materials. 2012;24:2547–2566.

[12] Oberlin A, Endo M, Koyama T. Journal of Crystal Growth. 1976,32:335.

[13] Iijima S. Nature. 1991;354:56.

[14] Iijima S, Ichihashi T. Nature. 1993;363:603.

[15] Bethune DS, Kiang CH, Devries MS, Gorman G, Savoy R, Vazquez J, Beyers R. Nature. 1993;363:605.

[16] Endo M, Muramatsu H, Hayashi T, Kim YA, Terrones M, Dresselhaus MS, Nature. 2005;433:476.

[17] Greiner A, Wendorff JH. Angewandte Chemie International Edition. 2007;46:5670.

[18] Teo WE, Inai R, Ramakrishna S. Science and Technology of Advanced Materials. 2011;12:013002.

[19] Chirayil CJ, Mathew L, Thomas S. Review of Advanced Materials Science. 2014;37:20–28.

[20] Visakh PM, Thomas S. Waste and Biomass Valorization. 2010;1:121.

[21] Rahman MM, Asiri AM, Youssef TE, Marwani HM. Material Express. 2016;6:137–148.

[22] Khan A, Khan AAP, Asiri AM, Rub MA, Rahman MM, Ghani SA. Microchimica Acta. 2014;181:1049.

Section 2

State-of-the-Art Biomaterials

Chapter 2

Nanofibers

Tayebeh Behzad and Mobina Ahmadi

Additional information is available at the end of the chapter

http://dx.doi.org/10.5772/63704

Abstract

Cellulose is the most available biopolymer on earth existent in wood and other lignocellulosic sources, which can be isolated as nanofibers to reinforce composites. The aim of this chapter overview was to explain cellulose nanofibers' preparation methods from different plant sources. The chemical treatment of cellulose fibers including dilute alkali swelling, acid and base hydrolysis, and bleaching was carried out followed by a mechanical method to isolate nanofibers. In this study, the effect of few palnt source variations and mechanical processes used in extraction procedure on nanofibers morphology, crystallinity, and chemical composition was evaluated. Characteristics of extracted cellulose nanofibers obtained from different plant sources were presented as well.

Keywords: natural fiber, extraction, cellulose nanofibers, chemical composition, morphology, crystallinity, thermal properties

1. Introduction

Over the past few decades, the use of plants as alternative sources of fibers instead of synthetic fibers has attracted more attention of many researchers. Employing natural fibers dates back to the early 1990s when German automakers were ahead in this field [1, 2]. The main goal of using natural fibers is recycling worthless materials which are found abundantly in nature. **Table 1** shows comparison between glass and and natural fibers properties [3].

Natural fiber application in construction and automotive industry is in significant progress due to low density and relatively low cost compared to glass fiber. Natural fibers are the miniature composites, which are composed of millions of microfibrils. Microfibrills are crystalline cellulose in a matrix contains lignin, hemicellulose , and other polysaccharides [4, 5]. According to the Food and Agricultural Organization (FAO) annually more than 35 million tons of fibers from plants and animals are produced in the world. These fibers have served useful purposes in the human society as they are widely used to make such valuable commodities as fabrics, ropes, papers, and packaging [6]. Lignocellulosic natural fibers obtained from plants are regarded as green, environmentally friendly materials due to their Biodegradability and high specific strength [7]. An important aspect of these fibers is their more recent application as reinforcement in composites. Compared to their inorganic counterparts, natural fibers used in composites offer such major benefits as abundance, low cost, renewability, and biodegradability, while they enjoy additional properties including low density (1.5 g/cm3) and large surface area (50–70 m^2/g) as well as such mechanical properties as high aspect ratio, modulus, and strength [8–10].

Properties	Natural fiber	Glass fiber
Density	Low	Double than natural fibers
Price	Low	Low
Renewable	Yes	No
Recycle	Yes	No
Biodegradability	Biodegradable	Nonbiodegradable

Table 1. Comparison between glass and natural fiber properties.

1.1. Sources of biofibers

Natural fibers, or lignocelluloses, are classified according to their wood and non-wood sources. Depending on the part of the plant that they are derived from, five categories of fibers are

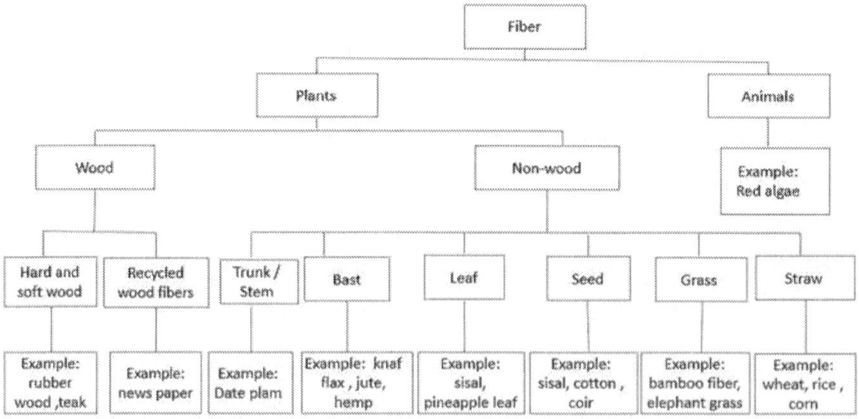

Figure 1. Classification of natural fibers.

http://dx.doi.org/10.5772/63704

identified: (1) bast or stem fibers; (2) leaf fibers; (3) seed-hair fibers, (4) core, pith, or straw fibers; and (5) all other plant fibers not included above. In other words, the four main categories of fibers are obtained from grain (cotton), vascular (flax, hemp, jute, rami, and wheat), grass (beet pulp, and bamboo), and algal sources (Valonica and Vetricosa).

Among the lignocellulosic sources, wood is consumed for paper production and cotton is used in the textile industry. However, only a small portion of annual plants such as hemp, wheat, soy hulls, bamboo, and other animal feed plants are used for energy production, most of which are burned as agricultural waste. **Figure 1** shows classification of natural fibers [8, 11, 12].

1.2. Structure of biofibers

Native fiber is comprised of small long thin filaments called microfibrils that consist of alternating crystalline and noncrystalline cellulose domains [13]. A single plant, typically 1–50 mm in length and around 10–50 μm in diameter, is called an elementary plant fiber. A plant fiber consist of few cells having cell walls compose of semicrystalline cellulose orientated in the direction of fiber and lying within a matrix of hemicellulose and lignin. A cell wall also surrounds a central lumen which help water uptake to the plant fiber [14, 15]. The fiber owes its high mechanical properties to 30–100 molecules of cellulose arranged in an extended chain conformation [16]. **Figure 2** presents an arrangement of cellulose fiber.

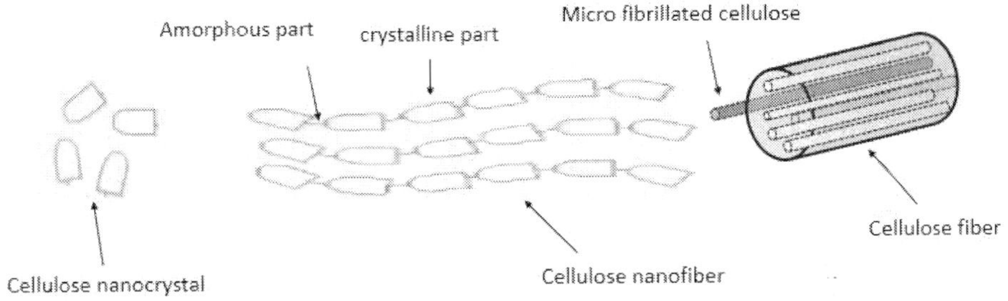

Figure 2. Scheme of cellulose fiber morphology.

Figure 3 shows the cell wall structure of a fiber [14, 17].

The secondary cell wall consists of an outer (S1), a middle (S2), and an inner (S3) layer, each consisting of microfibrils oriented in a specific angle with respect to one another [18–20].

The main constituents of natural fibers include cellulose, hemicellulose, lignin, pectin, wax, and water-soluble compounds. Hemicellulose is contains polysaccharides (excluding pectin) which remains connected to cellulose after lignin is removed. Hemicellulose is different from cellulose in three main ways [16]: First, it comprises various units of sugars (xylose, mannose, galactose, glucose, and rhamnose) [21]; second, it has a considerably higher degree of branching; and third, its degree of polymerization is 10–100 times lower than that of cellulose. In contrast to cellulose, the constituents of hemicellulose vary from plant to plant. Consequetly, hemicellulose is amorphous in structure and surrounds cellulose nanofibers [14, 16, 22].

Lignin is a perfectly amorphous and hydrophobic polymer formed by a combination of linear and cyclic hydrocarbon derivatives of phenyl propane. The structure of lignin is not yet known, but its mechanical properties are known to be inferior to those of cellulose [14, 16]. It provides the necessary hardness and firmness to plants. Lignin cannot be hydrolyzed in acid but is immediately oxidized in a warm alkaline environment and easily removed with phenol [16].

Pectin is the collective name for heteropolysaccharides mainly made of polygalacturonic acid. It is an amorphous polymer that gives flexibility to the plant [16].

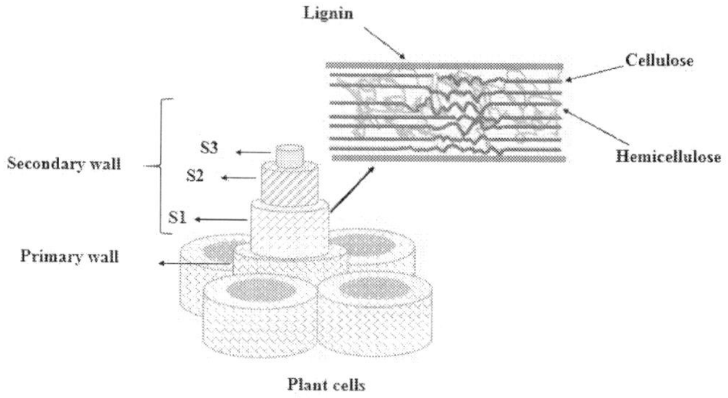

Figure 3. Scheme of plant fiber structure, S1, S2, and S3 layers.

1.3. Chemo-physical properties of cellulose

The specific structure of cellulose biopolymers defines their chemical and mechanical properties. As shown in **Figure 4**, the molecular structure of cellulose consists of β-D-anhydroglucopyranose repeating units that are bonded covalently together using acetal functional groups between the equatorial groups of C4 and C1 carbon atoms (β-1, 4-glycosidic bonds [23, 24].

Sensitivity to the hydrolytic attack of the β-1, 4-glycosidic linkages between the glucose repeating units is the factor determining the chemical stability of a cellulose molecule. It is clear from the molecular structure of cellulose that it is a linear polymeric chain having a large number of hydroxyl groups. The reason for this linear conformation is the β-glucose link at the C1–O–C4 to yield cellobiose units [23, 25, 26].

A second intramolecular hydrogen bonding may exist between the hydroxyl functional groups on the C6 and C2′ carbon atoms of the neighboring glucose units. Moreover, an intermolecular hydrogen bonding exists between the hydroxyl groups on the C6 and C3'' carbon atoms of cellulose molecules that are adjacently located on the same plane [23, 27–29].

http://dx.doi.org/10.5772/63704

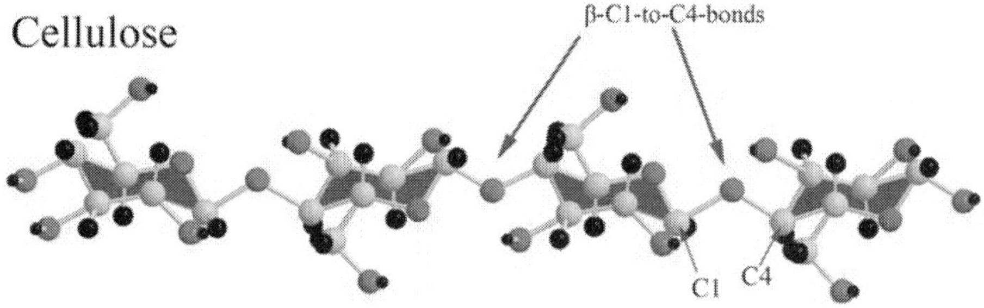

Figure 4. Show cellulose structure.

It is known from Infrared spectroscopy and X-ray diffraction (XRD) investigations of cellulose organization in plants that crystallites and low degree of scattered amorphous regions are the main constituents of cellulose [30, 31]. Native cellulose, namely cellulose I, is the crystalline cellulose while that precipitated out of alkali solutions is generally known as regenerated cellulose or cellulose II [30, 32]. Cellulose II is the most thermodynamically stable form due to the presence of an additional hydrogen bond per glucose residue [30, 33, 34].

It should be noted that cellulose gains its unique properties (**Table 2**) mentioned earlier after it reaches to its nanoscale [35].

Component	Young's modulus (MPa)
Biomass	20
Single fiber network	40
Microfibril	70
Cellulose nanocrystal	50–143
Cellulose nanofiber	145–150

Table 2. Young's modulus of cellulose after it reaches its nano-size [35].

Table 3 presents a comparison mechanical properties of glass, jute fibers and celluloe nanofibers [31, 36].

Specific fiber	Young's modulus (GPa)	Tensile strength (GPa)	Density (g/cm^3)
Glass fiber	~72.4 at 21°C	~3.4 at 21°C	2.54–2.62
Jute fiber	13–26	0.39–0.77	1.3–1.45
Cellulose nanofibers	~150	~10	1.5

Table 3. Comparison between mechanical properties of nanofibers and other fiber [31, 36].

1.4. Cellulose nanofibers (CNF) extraction methods

A variety of methods may be used for the extraction of cellulose nanofibers, each of which yields a unique type of nanofiber. Some of the extraction methods used are outlined below:

1.4.1. Biological method (hydrolytic enzyme)

This method is based on the use of micro-organisms, bacteria, or fungi. In this method, fibers are produced with diameters in the order of 100 nm form a bunch of nanofibers including those approximately 2 nm in diameter. These fibers have excellent intrinsic properties that yield crystallinity percentage values of around 89–84% [16].

1.4.2. Mechanical method

Mechanical or physical methods of extraction include cryocrushing, grinding, high-pressure homogenization, ultrasound, and steam explosion methods. The fibers obtained by these methods have a low strength and a small length-to-diameter ratio because of applying high tension on fibers. Another disadvantage of this method is its failure to easily manufacture fibers of nanosize diameters. [14, 16].

1.4.3. Chemo-mechanical method

These comprise the most common methods used for producing nanofibers with diameters less and length-to-diameter ratios higher than those produced by the mechanical method [16]. In these methods, the non-cellulosic content is removed by acid hydrolysis and alkali treatment [37]. Comparison of the advantages and disadvantages of conventional nanofiber extraction methods indicates that the chemical-mechanical ones are the safest for nanofiber isolation. **Figure 5** schematically present a chemo-mechanical method for production of cellulose nanofibers

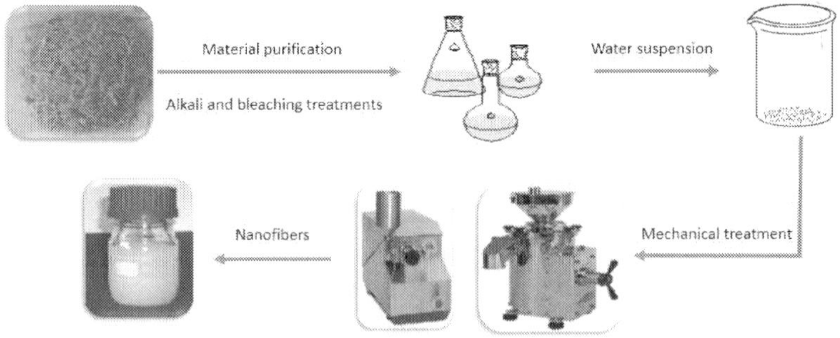

Figure 5. Scheme of chemo-mechanical method for production of cellulose nanofibers.

http://dx.doi.org/10.5772/63704

2. Experimental

2.1. Chemical treatment for CNF isolation

Some researchers have used an alkaline-acid pretreatment prior to mechanical isolation of CNF in order to solubilize lignin, hemicellulose, and pectins [2, 38]. The chemical treatment is accomplished in the following four steps:

(1) Soaking fibers in the 17.5 wt.% sodium hydroxide (NaOH) solution for 2 h; thus, the exposed surface area of the cellulosic fibers is enhanced and they become more susceptible to hydrolysis [16];

(2) Hydrolyzing the fibers in the hydrochloric acid (HCL) solution (2 M) at 80°C to solubilize hemicellulose [39];

(3) Treating the fibers with 2 wt.% NaOH solution for 2 h at 80°C in order to disrupt the lignin structure and to break down the carbohydrate and lignin linkages [16, 40]; and

(4) Bleaching the fibers with sodium hypochlorite, sodium chlorite, and hydrogen peroxide performed in an acidic environment at 50°C for 1 h [41].

2.2. Mechanical treatment

Cellulose nanofibers are extracted from the secondary cell walls by mechanical treatment without any damages to the cellulose. The degree of polymerization of cellulose, morphology, and aspect ratio of the CNFs depend on the defibrillation technique employed. Refining, grinding, and ultrasonication processes are performed for the mechanical isolation of CNF from wood and non-wood fibers [10, 42].

2.2.1. Grinding

Grinding is used to break up cellulose microfiber into nanosized fibers. The grinding machine consists of one rotating and one static grind stone where the pulp slurry passes through. In the fibrillation mechanism using a grinder, the hydrogen bonds are disintegrated under shear forces and the pulp is transformed into nanoscale fibers [42].

2.2.2. Ultrasonication

The mechanical process in which cellulose fibers are isolated using oscillating power by means of hydrodynamic forces of an ultrasound device is called "high intensity ultra sonication" (HIUS) [43]. When the ultrasonic energy is absorbed by the molecules, the cavitation evolving during the process give rises to a strong oscillating power, which generates high-intensity waves that lead to the formation, expansion, and explosion of microscopic gas bubbles [44]. Many efforts have been directed improving the HIUS resulting oscillating power to isolate nanofibers from cellulosic sources [42]. **Table 4** reports some of the plant sources of nanoscale cellulose fibers produced by a supergrinder and the ultrasonic system.

Process	Sources
Grinding and ultrasonic	Bagasse
Grinding and ultrasonic	
Ultrasonic	Rice straw
Grinding	Rice husk
Grinding and ultrasonic	Maize
Grinding and ultrasonic	Sawdust

Table 4. Different mechanical methods for isolating cellulosic nanofibers.

3. Results and discussions

3.1. Chemical composition characterization

An important factor affecting the application of natural fibers as biodegradable materials is their chemical composition. It is, therefore, essential to determine their chemical and mechanical properties of fibers used for manufacturing different products such as composites or newspaper. The three major chemical constituents of fibers extracted from lignocellulosic products are α-cellulose, lignin, and hemicellulose [45]. The factors affecting the properties of CNFs such as yield, dimensions, and mechanical properties of nanocellulose include chemical composition, the structure of the base material, and the processing method used for their extraction. Efficient isolation of cellulosic nanomaterials strongly depends on the removal of hemicellulose and lignin [1, 7]. Lignocellulosic fibers may be transformed into individual fibers during the extraction process by employing chemical purification via acid hydrolysis, alkali treatment, and bleaching. Applying the dilute alkali treatment to the lignocellulosic fibers breaks up the structural bonds between lignin and saccharides to remove the lignin content. Acid hydrolysis is used to destroy and solubilize the hemicellulose and pectin. Bleaching separates the Klasson (insoluble) lignin by oxidizers such as sodium chlorite, sodium hypochlorite, and hydrogen peroxide [2, 40, 46]. **Table 5** presents the chemical compositions of fibers extracted from different sources in the three raw, pre-, and post-bleaching stages. As it can be noticed the cellulose content from different sources after purification is in the range of 85 to 95 %; while the residual lignin content is obtained to be between 0.8 to 11%.

Figure 6 presents wheat-straw fibers before and after bleaching. Clearly, the brown color is present in the sample before the removal of insoluble lignin, indicating that lignin is an amorphous phase that keeps reducing with increasing removal efficiency. Comparison of the colors in this **Figure 5** shows that almost all the residual lignin (insoluble lignin) in the sample was eliminated and the sample becomes pure white after bleaching.

http://dx.doi.org/10.5772/63704

Sources	Cellulose (%)	Hemicellulose (%)	Lignin (%)
Bagasse			
Raw fiber	42.2	21.8	23.5
Before bleaching	88.23	8.82	2.95
After bleaching	91.8	7.37	0.82
Rice straw (stem)			
Raw fiber	46.5	22.5	29.1
Before bleaching	79.3	4.8	15.9
After bleaching	–	–	–
Rice husk			
Raw fiber	34.67	27.3	25.7
Before bleaching	67.36	14.83	11.74
After bleaching	94.3	1.8	1.42
Wheat straw			
Raw fiber	50.01	25.32	24.46
Before bleaching	86.2	3.7	10.1
After bleaching	93.7	2.95	3.35
Wheat straw (stem)			
Raw fiber	49.8	27.2	23.1
Before bleaching	85.5	3.8	10.8
After bleaching	–	–	–
Sawdust			
Raw fiber	33.38	28.71	37.08
Before bleaching	57.84	18.71	23.2
After bleaching	86.24	7.76	5.98

Table 5. Chemical compositions of fibers from different sources after selective chemical treatments.

Raw fiber Before bleaching After bleaching

Figure 6. Image of raw wheat straw fiber, and its pulp before and after bleaching.

3.2. Morphology description

Cellulosic fibers are made of cell walls that collectively form its structure. It is commonly established that chemical and mechanical treatments bring about changes in fiber morphology [10, 47]. Microscopic examinations indicate that raw fiber bundles consist of individual microfibers that are bonded together by lignin (**Figure 6**). Hence, chemical treatments are used to reduce the surface roughness and the size of the bundles. It has been reported that fiber diameter is reduced to the nanoscale as a result of chemical and mechanical treatments. This can be related to the removal of such non-cellulose constituents as lignin, hemicellulose, and waxes.

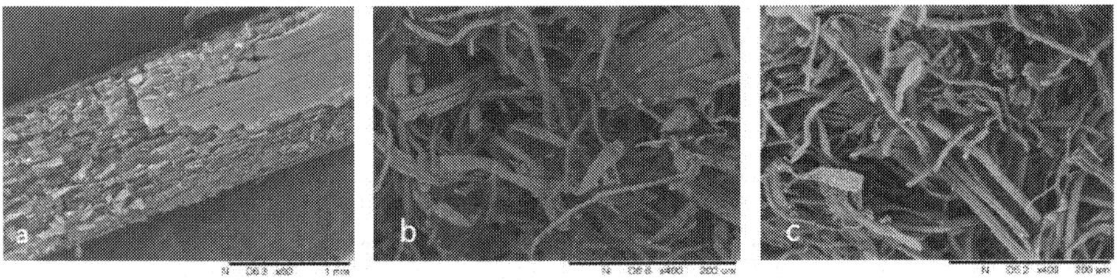

Figure 7. SEM images of a) raw fiber, b) pre-treated fiber, and c) bleached fiber.

Scanning electron microscopy (SEM), field-emission scanning electron microscopy (FE-SEM), transmission electron microscopy (TEM), and atomic force microscopy (AFM) techniques are used to characterize the morphology of the nanocellulose fibers produced. As can be seen, in **Figure 6**, the microfiber diameter is reduced as a result of chemical operations due to removal of the non-cellulosic material. It is clear from the the FE-SEM and TEM images that the extracted nanofibers have a network structure and are in the form of entangled fibrils. **Table 6** presents the variations in the diameter of the nanocellulose fibers isolated from different plant sources. The CNFs can have diameters varying from 15–200 nm depending on their source and the mechanical method used to disintegrate them. The average diameters of individual CNFs vary from 18 to 45 nm.

Sources	Process	Diameter Average (nm)
Bagasse	Ultrasonic 700 W	44
Rice straw	Grinding	19
Wheat straw	Grinding	27.89
Wheat straw	Ultrasonic 400 W	45
Sawdust	Grinding	21.74
Sawdust	Ultrasonic 400 W	1–15 μm

Table 6. Diameters and aspect ratios of cellulosic nanofibers extracted from different sources.

Comparing the two mechanical methods used to convert microfibers to nanofibers, the grinder method was found to have a higher mechanical efficiency. This is due to its higher power compared to that of the ultrasonic device. It was found (**Table 6**) the fiber extracted from sawdust disintegrated by ultrasonic 400 have diameter between 1-15 μm. The reasons for this could be: First, the low shear stress of the ultrasound to separate the fibrils; and second, the stronger fiber structure of sawdust compared to the other sources of lignocellulose listed in **Table 6**.

Figure 8 shows the FE-SEM and TEM of CNFs as a network of intertwined long nanofibers with a cellulosic appearance.

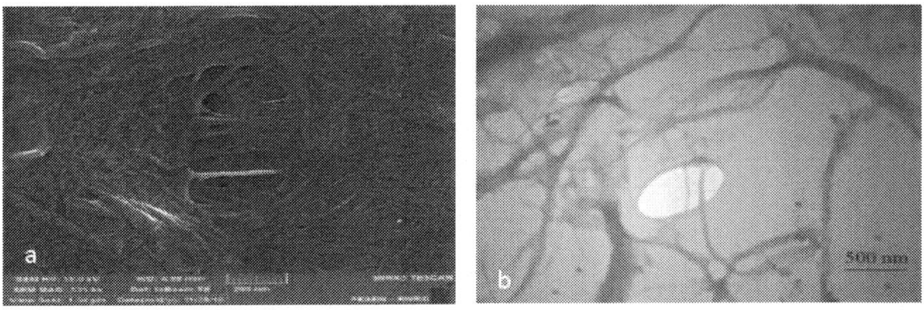

Figure 8. (a) FE-SEM and (b) TEM of cellulose nanofibers.

3.3. Crystallinity and thermal properties

The amount of crystallinity in the cellulosic nanofibers is expressed as total crystallinity index (TCI). Represented by Equation (1) in this equation, A_{1372} is absorption in 1372 cm^{-1} and A_{2900} is absorbtion in 2900 cm^{-1} in ATR spectrum. Generally, the peak in the infrared spectrum at 1372 cm^{-1} represents the modified C-H standardization of crystalline regions, and reveals any change in crystallinity, and that at 2900 cm^{-1} represents C-H stretching, which is a measure of amorphous regions. The absorption band at these two points represents the total crystallinity index, which is representative of the degree of crystallinity. The TCI values are reported in **Table 7**.

$$TCI = \frac{A_{1372}}{A_{2900}} \tag{1}$$

One well-known characteristic of cellulosic materials is that they experience fast thermal degradation at low to moderate temperatures, particularly at temperatures below 400°C. The thermal decomposition of lignocellulosic materials begins with the disintegration of hemicellulose followed by pyrolysis of lignin, depolymerization of cellulose, active flaming combustion, and char oxidation. The thermal stability of cellulosic fibers is an important parameter for using them in the manufacture of composites. The structure and chemical composition of cellulosic fibers drastically influence their thermal Degradation. The first stage takes place at temperatures ranging from room temperature up to 120°C, at which a little weight loss is observed in the TGA thermogram of CNFs; Which is related to the evaporation of water. The second stage Occurres at temperatures in the range of 220 to 390°C, where the destruction of the crystalline regions are started and polymer decomposition is simultaneously triggered, which led to enhance the amorphous structure and ultimately to reduction in the degree of polymerization. The third stage from 390 to 600°C involves the complete destruction of the crystalline region and the formation of D-glucopyranosemonomers by the decomposition of cellulose, which could be further converted into free radicals [48–50].

Sources	TCI
Bagasse	1.26
Rice straw	1.4
Rice husk	1.56
Wheat straw	1.32
Wheat straw	1.232
Maize	1.564
Sawdust	2.507

Table 7. TCI values of nanofibers from different sources.

4. Conclusions

Today, cellulose nanofibers because of their unique reinforcing effect are widely used in Nanocomposite. The effect of type of plant source and mechanical process used for production of nanofibers, on its properties, including the specific surface area and aspect ratio is impressive. Among the mechanical methods used to produce nanofibers, super grinder has the maximum efficiency. Due to creating higher tension. Different sources,produce nanofibers with cellulose content ranging between 85–95% and diameter changing between 18–45 nm depend on the employed mechanical method.

Author details

Tayebeh Behzad* and Mobina Ahmadi

*Address all correspondence to: tbehzad@cc.iut.ac.ir

Department of Chemical Engineering, Isfahan University of Technology, Isfahan, Iran

References

[1] Chirayil, C.J., L. Mathew, and S. Thomas:Review of recent research in nano cellulose preparation from different lignocellulosic fibers. Reviews on Advanced Materials Science. 2014. 37: 20–28.

[2] Alemdar, A. and M. Sain:Isolation and characterization of nanofibers from agricultural residues – Wheat straw and soy hulls. Bioresource Technology. 2008. 99(6) 1664–1671. doi: http://dx.doi.org/10.1016/j.biortech.2007.04.029

[3] Wambua, P., J. Ivens, and I. Verpoest:Natural fibres: Can they replace glass in fibre reinforced plastics? Composites Science and Technology. 2003. 63(9): 1259–1264. doi: http://dx.doi.org/10.1016/S0266-3538(03)00096-4

[4] Fowler, P.A., J.M. Hughes, and R.M. Elias:Biocomposites: Technology, environmental credentials and market forces. Journal of the Science of Food and Agriculture. 2006. 86(12): 1781–1789. doi: 10.1002/jsfa.2558

[5] Fowler, P.A., J.M. Hughes, and R.M. Elias:Biocomposites from crop fibres and resins. IGER Innovations. 2007: 66–68.

[6] FAO. Available online: www.naturalfibres 2009.org (accessed on 8 December 2010).

[7] Bledzki, A.K. and J. Gassan:Composites reinforced with cellulose based fibres. Progress in Polymer Science. 1999. 24(2): 221–274. doi: http://dx.doi.org/10.1016/S0079-6700(98)00018-5

[8] Mohanty, A.K., M. Misra, and L.T. Drzal:Sustainable bio-composites from renewable resources: Opportunities and challenges in the green materials world. Journal of Polymers and the Environment. 10(1): 19–26. doi: 10.1023/a:1021013921916

[9] Mohanty, A.K., M. Misra, and G. Hinrichsen:Biofibres, biodegradable polymers and biocomposites: An overview. Macromolecular Materials and Engineering. 2000. 276–277(1): 1–24. doi: 10.1002/(SICI)1439-2054(20000301)276:1<1::AID-MAME1>3.0.CO;2-W

[10] Jonoobi, M., et al.:Different preparation methods and properties of nanostructured cellulose from various natural resources and residues: A review. Cellulose. 2015. 22(2): 935–969. doi: 10.1007/s10570-015-0551-0

[11] Abdul Khalil, H.P.S., et al.:Conventional agro-composites from chemically modified fibres. Industrial Crops and Products. 2007. 26(3): 315–323. doi: http://dx.doi.org/10.1016/j.indcrop.2007.03.010

[12] Akil, H.M., et al.:Kenaf fiber reinforced composites: A review. Materials & Design. 2011. 32(8–9): 4107–4121. doi: http://dx.doi.org/10.1016/j.matdes.2011.04.008

[13] Dufresne, A., et al.:Biopolymer Nanocomposites: Processing, Properties, and Applications. Wiley. 2013. ISBN: 978-1-118-21835-8 Editor: Alain Dufresne, Sabu Thomas ,Laly A. Pothan , DOI: 10.1002/9781118609958

[14] Lavoine, N., et al.:Microfibrillated cellulose – Its barrier properties and applications in cellulosic materials: A review. Carbohydrate Polymers. 2012. 90(2): 735–764. doi: http://dx.doi.org/10.1016/j.carbpol.2012.05.026

[15] Wang, B. and M. Sain:Isolation of nanofibers from soybean source and their reinforcing capability on synthetic polymers. Composites Science and Technology. 2007. 67(11–12): 2521-2527. doi: http://dx.doi.org/10.1016/j.compscitech.2006.12.015

[16] Thomas, S., et al.:Natural fibres: Structure, properties and applications, in: Cellulose Fibers: Bio- and Nano-Polymer Composites: Green Chemistry and Technology, S. Kalia, S.B. Kaith, and I. Kaur, Editors, Springer: Berlin, Heidelberg. 2011. pp. 3–42.

[17] Rong, M.Z., et al.:The effect of fiber treatment on the mechanical properties of unidirectional sisal-reinforced epoxy composites. Composites Science and Technology. 2001. 61(10): 1437–1447. doi: http://dx.doi.org/10.1016/S0266-3538(01)00046-X

[18] Geethamma, V.G., et al.:Composite of short coir fibres and natural rubber: Effect of chemical modification, loading and orientation of fibre. Polymer. 1998. 39(6–7): 1483–1491. doi: http://dx.doi.org/10.1016/S0032-3861(97)00422-9

[19] Mukherjee, P.S. and K.G. Satyanarayana:An empirical evaluation of structure-property relationships in natural fibres and their fracture behaviour. Journal of Materials Science. 21(12): 4162–4168. doi: 10.1007/bf01106524

[20] Mahato, D.N., B.K. Mathur, and S. Bhattacherjee:Effects of alkali treatment on electrical and spectral properties of coir fibre. Journal of Materials Science Letters. 12(17): 11350–1353. doi: 10.1007/bf00241705

[21] Mohammad Tajul Islam, M.M.A., Marina Zoccola:Review on modification of nanocellullose for application in composites. International Journal of Innovative Research in Science Engineering and Technology. 2013. 2(10) 5444–5451.

[22] Kumar, A.P. and R.P. Singh:Biocomposites of cellulose reinforced starch: Improvement of properties by photo-induced crosslinking. Bioresource Technology. 2008. 99(18): 8803–8809. doi: http://dx.doi.org/10.1016/j.biortech.2008.04.045

[23] Kennedy, J.F. and W.H.W. Hasamudin:Cellulose: Structure, accessibility and reactivity, in: Polymer International, H. A. Krassig, Editor, Gordon & Breach, Pennsylvania, 1993. Pp. xvi + 376, price £169.00, US$260.00. ISBN 2-88124-798-9. 1995. 36(1): 101–101. doi: 10.1002/pi.1995.210360114

[24] Kennedy, J.F.:Cellulose and its Derivatives: Chemistry, Biochemistry, and Applications. E. Horwood. 1985. ISBN 0470201975, 9780470201978

[25] Ng, H.-M., et al.:Extraction of cellulose nanocrystals from plant sources for application as reinforcing agent in polymers. Composites Part B: Engineering. 2015. 75: 176–200. doi: http://dx.doi.org/10.1016/j.compositesb.2015.01.008

[26] Klemm, D., et al.:Cellulose: Fascinating biopolymer and sustainable raw material. Angewandte Chemie International Edition. 2005. 44(22): 3358–3393. doi: 10.1002/anie.200460587

[27] Andresen, M., et al.:Properties and characterization of hydrophobized microfibrillated cellulose. Cellulose. 2006. 13(6): 665–677. doi: 10.1007/s10570-006-9072-1

[28] Dufresne, A., J.-Y. Cavaillé, and M.R. Vignon:Mechanical behavior of sheets prepared from sugar beet cellulose microfibrils. Journal of Applied Polymer Science. 1997. 64(6): 1185–1194. doi: 10.1002/(SICI)1097-4628(19970509)64:6<1185::AID-APP19>3.0.CO;2-V

[29] Stenstad, P., et al.:Chemical surface modifications of microfibrillated cellulose. Cellulose. 2007. 15(1): 35–45. doi: 10.1007/s10570-007-9143-y

[30] Mark, H.F. and J.I. Kroschwitz. Encyclopedia of Polymer Science and Engineering. Wiley. 1985.

[31] Siqueira, G., J. Bras, and A. Dufresne:Cellulosic bionanocomposites: A review of preparation, properties and applications. Polymers. 2010. 2(4): 728.

[32] Dufresne, A.:Polymer nanocomposites from biological sources. Encyclopedia of Nanoscience and Nanotechnolog. 2010: 1–23. ASP-ENN-08-0301/Application/ASP-ENN-08-0301-004.3d.

[33] Sugiyama, J., J. Persson, and H. Chanzy:Combined infrared and electron diffraction study of the polymorphism of native celluloses. Macromolecules. 1991. 24(9): 2461–2466. doi: 10.1021/ma00009a050

[34] Saxena, I.M. and R.M. Brown:Cellulose biosynthesis: Current views and evolving concepts. Annals of Botany. 2005. 96(1): 9–21. doi: 10.1093/aob/mci155

[35] Rosa, M.F., et al.:Cellulose nanowhiskers from coconut husk fibers: Effect of preparation conditions on their thermal and morphological behavior. Carbohydrate Polymers. 2010. 81(1): 83–92. doi: http://dx.doi.org/10.1016/j.carbpol.2010.01.059

[36] Hu, W., et al.:Functionalized bacterial cellulose derivatives and nanocomposites. Carbohydrate Polymers. 2014. 101: 1043–1060. doi: http://dx.doi.org/10.1016/j.carbpol.2013.09.102

[37] Zhang, J., et al.:Facile synthesis of spherical cellulose nanoparticles. Carbohydrate Polymers. 2007. 69(3): 607–611. doi: http://dx.doi.org/10.1016/j.carbpol.2007.01.019

[38] Chen, W., et al.:Isolation and characterization of cellulose nanofibers from four plant cellulose fibers using a chemical-ultrasonic process. Cellulose. 2011. 18(2): 433–442. doi: 10.1007/s10570-011-9497-z

[39] Alidadi-Shamsabadi, M., et al.:Preparation and characterization of low-density polyethylene/thermoplastic starch composites reinforced by cellulose nanofibers. Polymer Composites. 2015. 36(12): 2309–2316. doi: 10.1002/pc.23144

[40] Komuraiah, A., N.S. Kumar, and B.D. Prasad:Chemical composition of natural fibers and its influence on their mechanical properties. Mechanics of Composite Materials. 2014. 50(3): 359–376. doi: 10.1007/s11029-014-9422-2

[41] Shen, W. and X. Chen:Measuring and controlling model of pulp kappa number with spectroscopy during batch sulfite pulping process. Industrial & Engineering Chemistry Research. 2009. 48(19): 8980–8984. doi: 10.1021/ie900327a

[42] Abdul Khalil, H.P.S., et al.:Production and modification of nanofibrillated cellulose using various mechanical processes: A review. Carbohydrate Polymers. 2014. 99: 649–665. doi: http://dx.doi.org/10.1016/j.carbpol.2013.08.069

[43] Cheng, Q., S. Wang, and T.G. Rials:Poly(vinyl alcohol) nanocomposites reinforced with cellulose fibrils isolated by high intensity ultrasonication. Composites Part A: Applied Science and Manufacturing. 2009. 40(2): 218–224. doi: http://dx.doi.org/10.1016/j.compositesa.2008.11.009

[44] Chen, P., et al.:Concentration effects on the isolation and dynamic rheological behavior of cellulose nanofibers via ultrasonic processing. Cellulose. 2012. 20(1): 149–157. doi: 10.1007/s10570-012-9829-7

[45] Sun, X.F., et al.:Degradation of wheat straw lignin and hemicellulosic polymers by a totally chlorine-free method. Polymer Degradation and Stability. 2004. 83(1): 47–57. doi: http://dx.doi.org/10.1016/S0141-3910(03)00205-2

[46] Crônier, D., B. Monties, and B. Chabbert:Structure and chemical composition of bast fibers isolated from developing hemp stem. Journal of Agricultural and Food Chemistry. 2005. 53(21): 8279–8289. doi: 10.1021/jf051253k

[47] Meeting, A.C.S., et al.:Cellulose Nanocomposites: Processing, Characterization, and Properties. American Chemical Society. 2006.

[48] Hu, W., et al.:Solvent-free acetylation of bacterial cellulose under moderate conditions. Carbohydrate Polymers. 2011. 83(4): 1575–1581. doi: http://dx.doi.org/10.1016/j.carbpol.2010.10.016

[49] Yang, Z., et al.:Characterization and acetylation behavior of bamboo pulp. Wood Science and Technology. 2008. 42(8): 621–632. doi: 10.1007/s00226-008-0194-5

[50] Ashori, A., et al.:Solvent-free acetylation of cellulose nanofibers for improving compatibility and dispersion. Carbohydrate Polymers. 2014. 102: 369–375. doi: http://dx.doi.org/10.1016/j.carbpol.2013.11.067

Chapter 3

Nanofibrous Scaffolds as Promising Cell Carriers for Tissue Engineering

Lucie Bacakova, Marketa Bacakova, Julia Pajorova, Radmila Kudlackova, Lubica Stankova, Elena Filova, Jana Musilkova, Stepan Potocky and Alexander Kromka

Additional information is available at the end of the chapter

http://dx.doi.org/10.5772/63707

Abstract

Nanofibers are promising cell carriers for tissue engineering of a variety of tissues and organs in the human organism. They have been experimentally used for reconstruction of tissues of cardiovascular, respiratory, digestive, urinary, nervous and musculoskeletal systems. Nanofibers are also promising for drug and gene delivery, construction of biosensors and biostimulators, and wound dressings. Nanofibers can be created from a wide range of natural polymers or synthetic biostable and biodegradable polymers. For hard tissue engineering, polymeric nanofibers can be reinforced with various ceramic, metal-based or carbon-based nanoparticles, or created directly from hard materials. The nanofibrous scaffolds can be loaded with various bioactive molecules, such as growth, differentiation and angiogenic factors, or funcionalized with ligands for the cell adhesion receptors. This review also includes our experience in skin tissue engineering using nanofibers fabricated from polycaprolactone and its copolymer with polylactide, cellulose acetate, and particularly from polylactide nanofibers modified by plasma activation and fibrin coating. In addition, we studied the interaction of human bone-derived cells with nanofibrous scaffolds loaded with hydroxyapatite or diamond nanoparticles. We also created novel nanofibers based on diamond deposition on a SiO_2 template, and tested their effects on the adhesion, viability and growth of human vascular endothelial cells.

Keywords: nanofibers, nanoparticles, natural polymers, synthetic polymers, ceramics, carbon, diamond, biomaterial, biocompatibility, tissue engineering, tissue regeneration, nanomedicine, drug delivery, gene delivery, wound healing

1. Introduction

In recent years, nanofibrous materials are becoming more and more popular for tissue engineering applications, because they mimic nanofibrous components of the native extracellular matrix (ECM), for example, collagen fibers. Nanofibers are also widely used in other biotechnologies, such as drug and gene delivery [1–5], gene silencing [6], construction of biosensors [7, 8], or preparation of wound dressings absorbing the exudate and protective against microbial infection [9, 10].

Nanofibers are typically prepared from polymeric materials, such as natural and synthetic polymers and their various combinations. Nature-derived polymers comprise a wide range of proteins, peptides, and polysaccharides, for example collagen [11, 12], elastin [13] and elastin-like peptides [14], silk fibroin [15, 16], amyloid [3], chitosan [12, 17], cellulose [10, 18], glycosaminoglycans [11, 12], or hyaluronan [19, 20]. Even demineralized bone matrix (DBM) was used for preparation of nanofibers by elecrospinning [21]. In fact, DBM, a natural polymer, is allograft bone with inorganic material removed. DBM contains the protein components of bone, which includes adhesion ligands and osteoinductive signals, such as important growth factors. The DBM nanofiber mats exhibited good cytocompatibility with human dermal fibroblasts [21].

Synthetic polymers include a broad spectrum of biostable polymers, for example polyethylene terephthalate [9], polytetrafluoroethylene [22] and polyurethane, suitable for fabrication of vascular grafts [23], and biodegradable polymers, for example polylactides (PLA) [24, 25] and their copolymers with polyglycolides (PLGA) [1, 26], polycaprolactone (PCL) [2, 4, 6], poly-3-hydroxybutyrate-co-3-hydroxyvalerate (PHBV) [27], polydioxanone [5], polyvinylalcohol (PVA) [28, 29], and synthetic peptides [7, 30].

However, pure polymeric nanofibers are suitable mainly for soft tissue engineering, such as reconstruction and regeneration of blood vessels [13, 22, 23, 31, 32], myocardium [33, 34], heart valves [35, 36], skeletal muscle [37, 38], skin [15, 39–41], tendon and ligament [42, 43], intestine [44, 45], tissues of the respiratory system, such as trachea and bronchi [46, 47], components of urinary tract, such as bladder [48] and urethra [49], visceral organs, such as liver [50, 51] or pancreas (pancreatic islets [52, 53]), central nervous system, such as brain [6, 54, 55], spinal cord [56, 57], optic system, such as optical nerve [58] and retina [59], and peripheral nervous system [17, 60]. Nanofibrous scaffolds can be associated with another advanced technique in recent tissue engineering—controlled delivery of various types of stem cells, such as bone marrow mesenchymal stem cells [51, 61–63], adipose tissue-derived stem cells [64, 65], neural tissue-derived stem cells [57], and induced pluripotent stem cells [20, 34], and their appropriate differentiation into desired cell types.

For hard tissue engineering, that is for reconstruction of bone, teeth, cartilage, and osteochondral interface, it is necessary to improve mechanical properties of the nanofibers. This is feasible by addition of inorganic nanoparticles into nanofibers, such as ceramic nanoparticles, for example hydroxyapatite [16, 27, 28, 30], tricalcium phosphate [63, 66], calcium oxide [67], or calcium silicate [24]; metal-based nanoparticles, for example gold nanoparticles [29], ferromagnetic Fe_3O_4 nanoparticles [25], or antimicrobial silver nanoparticles [26]; and also carbon-based nanoparticles, such as carbon nanotubes, graphene [62], or nanodiamonds [68]. These nanoparticles not only reinforce the polymeric nanofibers but also enhance their bioactivity in terms of increased cell adhesion, growth, osteogenic cell differentiation, bone matrix mineralization and antimicrobial activity. The mineral component can also be added to the nanofibers by biomimetic mineralization in simulated body fluid [69] and other ionic solutions [70]. Nanofibers can also be created exclusively from inorganic or other hard materials, for example hydroxyapatite [71, 72] or carbon and bioactive glass [73], and also from SiO_2 [74, 75] and diamond [76, 77] or their combinations [78, 79].

On the other hand, carbon nanoparticles (mainly carbon nanotubes) have been also added into nanofibers for soft tissue engineering in order to allow electrical stimulation of cells or delivery of drug and other bioactive molecules to the cells. The electrical stimulation is suitable especially for cells of excitable tissues, such as neural tissue [80] (for a review, see [81]), myocardium [82], skeletal muscle [38], and vascular smooth muscle [83]. The addition of carbon nanoparticles also improved the mechanical properties of the nanofibrous scaffolds for engineering of muscular tissues, which are exposed to a relatively high mechanical loading in the organism. For the purpose of electrical stimulation of cells, nanofibers can be also coated with polypyrrole [55] or directly made of this polymer [58].

Nanofibers can be further loaded with various biomolecules in order to achieve their specific effects on cells, for example, with growth factors such as basic fibroblast growth factor and epidermal growth factor [12, 40, 42, 84], with angiogenic factors, such as vascular endothelial growth factor or platelet-derived growth factor [52, 85], with differentiation factors, such as bone morphogenetic protein-2 (BMP-2) [18] or with brain-derived neurotrophic factor [54]. Other bioactive agents include ascorbic acid, promoting the production of collagen by cells [86, 87], glutamate for neural tissue engineering [56], vitamin E and polyphenols (e.g., curcumin and green-tea polyphenols), that are natural compounds with excellent antioxidant, anticancer, antimicrobial, anti-inflammatory and wound-healing properties [88, 89], hormones and their analogues (estradiol [10], dexamethasone [89]), honey [9], and propolis [90]. The cell adhesion and growth on nanofibrous scaffolds can be enhanced by their functionalization with ligands for cell adhesion receptors, for example RGD-containing oligopeptides [20, 30, 39].

Nanofibers can be prepared by various techniques, for example self-assembly (silk-elastin-like protein polymers [14]), interfacial polymerization, suitable for electrically conductive materials [8], melt processing [91], and antisolvent precipitation [92]. The latter two methods are suitable for preparation of porous nanofibers for loading various substances. However, the most effective method for large-scale production of nanofibers is electrospinning, particularly needleless electrospinning [93]. Polymer composites produced via the needleless electrospinning allow a polymer nanofiber to act as a host for nanoparticles, and, in addition, the polymer nanofibers will act as a three-dimensional carrier for cells imitating natural extracellular matrix.

This chapter comprises our experience in using nanofibers for experimental soft and hard tissue engineering in correlation with studies of other authors in recent years. We have focused particularly on *skin tissue engineering* using polymeric nanofibers made of PLA, PCL, PLA/PCL or cellulose, and further modified with plasma or coated with fibrin and on *bone tissue engineering* using polymeric nanofibers loaded with hydroxyapatite or diamond nanoparticles, or SiO_2 nanofibers coated with nanodiamond.

2. Nanofibers in skin tissue engineering

The skin is composed of three main layers—epidermis, dermis, and hypodermis. The epidermis, the outermost layer, consists mainly of keratinocytes (more than 90% of all cell types) but it also contains subpopulations of melanocytes, Langerhans cells, and Merkel cells. The keratinocytes produce many important molecules, such as growth factor and protective immunogenic molecules. These molecules include interleukins, transforming growth factors α and β, platelet-derived growth factor, fibroblast growth factor, tumor necrosis factor α, and interferons α and β. The melanocytes produce the pigment melanin that protects the skin against harmful effect of sunlight. The Langerhans cells, that is a type of leucocytes, are responsible for immune activation. The function of Merkel cells is not yet clearly elucidated yet but they probably occasionally participate in formation of synaptic junction with peripheral nerves, and in low-vertebrates, they participate in slow-adapting touch perception [94].

The epidermal cells form five sublayers: *stratum basale, stratum spinosum, stratum granulosum, stratum lucidum,* and *stratum corneum*. New cells are created in the deepest basal layer as stem cells, and then, they differentiate and mature in adult keratinocytes and move towards the skin surface. As keratinocytes mature and ascend toward the epidermal layers, their shapes become flattened, and these cells synthetize structural protein keratin. The outermost cornified layer is created by dead keratinocytes rich in protein keratin. The importance of keratinization is creating a barrier to prevent fluid loss and unwanted entry of potentially harmful molecules and microorganisms.

Dermis, situated below the epidermis, is responsible for elasticity and mechanical integrity of the skin, cutaneous nutrition, immunosurveillance, sensory perception, and temperature regulation. The main cellular type of dermis are fibroblasts that are responsible for synthetic and degradation of dermal proteins. The other cells included in the dermis are endothelial cells, smooth muscle cells and immune cells (dendritic cells, monocytes, and lymphocytes). The dermis also contains nerves, vessels, sweet glands, and hair follicles [94].

Hypodermis, the undermost layer, is mainly composed of adipose tissue and collagen and acts as a fat storage, an energy source, and enables an anchorage of the skin to bone or muscle [94].

Skin tissue engineering is focused mainly on the reconstruction of epidermis and dermis using a biomaterial scaffold (as cell carrier) and two main epidermal and dermal cell types, that are keratinocytes and fibroblasts, respectively. Nanofibrous meshes can be advantageously used for creating a bilayered epidermal–dermal construct containing keratinocytes and fibroblasts located on the opposite sides of the membrane. The keratinocytes and fibroblasts could communicate physically and biochemically through the pores in the membrane, if the membrane is of appropriate thickness and pore size. The crosstalk between the keratinocytes and fibroblasts contributed to epidermal stratification, higher tensile strength of the construct, modulation of cytokine and growth factor expression, and increased angiogenic properties compared with constructs containing fibroblasts or keratinocytes alone [95]. Nevertheless, the bilayered epidermal–dermal construct has been rarely developed on nanofibrous membranes. For its creation, other forms of synthetic and natural polymers have been used, for example a knitted PLGA mesh combined with collagen–hyaluronic acid sponge [96], porous scaffolds made of a copolymer of poly(ethylene glycol terephthalate) and poly(butylene terephthalate) [97] matrices containing fibrin, collagen, hyaluronan and their combinations [98–100], and even spider silk woven on steel frames [101]. Thus, we have attempted to develop a bilayer of keratinocytes and skin fibroblasts using nanofibrous PLA membranes, fixed in adapted CellCrown inserts (Scaffdex, Tampere, Finland). The membranes were first seeded with fibroblasts, because these cells served as a feeder layer for keratinocytes, and after reaching confluence of fibroblasts (on day 7 after seeding), the inserts were converted, seeded with keratinocytes, and these cells were cultivated for 4 days (**Figure 1**).

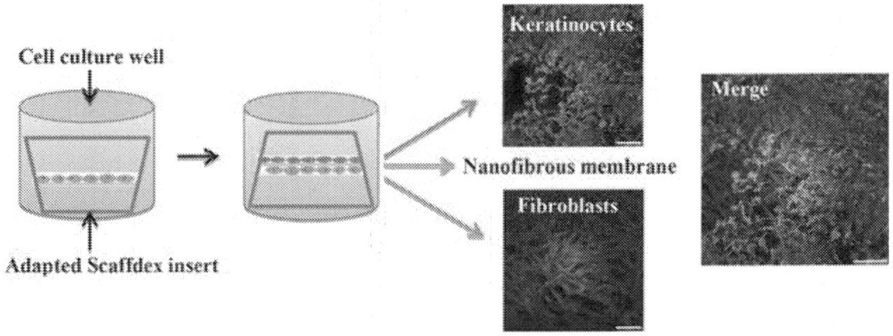

Figure 1. Development of a bilayered construct of HaCaT keratinocytes and neonatal human dermal fibroblasts on the opposite sides of a nanofibrous membrane (scale bar = 100 μm). The keratinocytes were stained by immunofluorescence against cytokeratin 5, an early marker of keratinocyte differentiation. The fibroblasts were stained with Texas Red C_2-Maleimide and Hoechst #33258.

As evident from the **Figure 1**, the keratinocyte layer on the PLA nanofibrous membrane was not continuous. In general, synthetic polymeric materials in their pristine state are rather hydrophobic and may behave as bioinert. Thus, we activated the PLA membranes by treatment with oxygen plasma in order to improve the cell-material interaction. We found that the plasma treatment improved the adhesion and growth of human HaCaT keratinocytes, which was manifested by the formation of larger cell islands (**Figure 2**), and also by an increased activity of mitochondrial enzymes (measured by the XTT assay), and increased DNA content (measured by the Picogreen dsDNA assay kit), which both are indicators of an increased cell number [102]. These beneficial effects of the plasma treatment of the cell behavior could be attributed to the formation of new oxidized structures on the membrane surface, increase in surface wettability, and changes in surface stiffness. Higher plasma power and, in particular, longer exposure times resulted in more pronounced improvement of the cell adhesion and growth. The fiber density of the membranes also played an important role in cell adhesion and growth. The cells preferentially adhered on membranes of lower fiber densities, due to the larger void spaces between the fibers. Thus, PLA nanofibrous membranes subjected to physical modifications proved as promising materials for the construction of temporary carriers for skin cells [102].

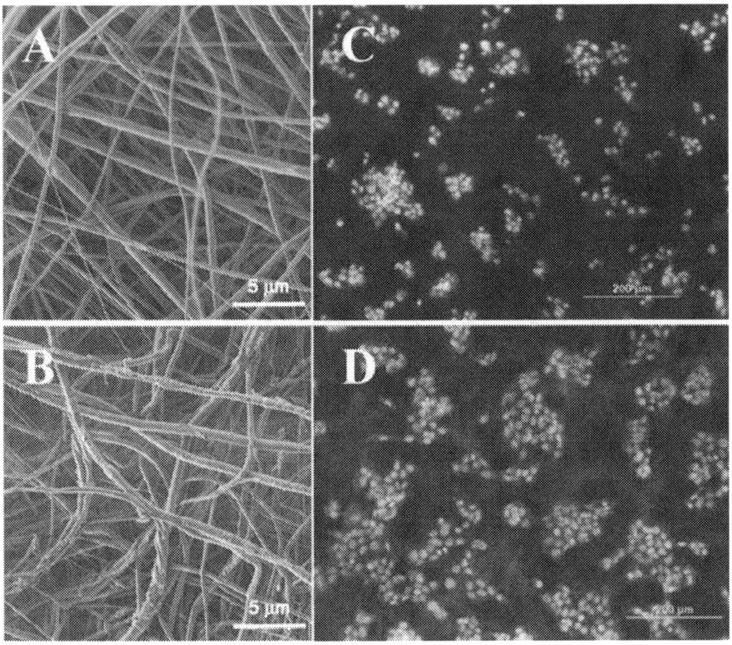

Figure 2. The morphology of unmodified nanofibrous polylactide scaffolds (A) and scaffolds treated by oxygen plasma (power 75 W, time 30 min, B), and the morphology of human HaCaT keratinocytes on these scaffolds (C and D). Note larger keratinocyte islands on plasma-treated scaffolds (D) than on untreated scaffolds (C). A, B: FE-SEM Tescan MIRA3 scanning electron microscope, objective magnification 10,000×, scale bar 5 µm. C, D: Cells stained with Texas Red C_2-Maleimide and Hoechst #33342. Olympus IX 51 microscope, obj. 10×, DP 70 digital camera, scale bar = 200 µm. Day 3 after seeding.

Also the modification of polymeric membranes with fibrin, that is a provisional matrix molecule playing an important role in tissue regeneration, had beneficial effects on the adhesion, growth, and functioning of skin cells, particularly human dermal fibroblasts (**Figure 3**). Fibrin films were developed by *in vitro* simulation of a specific part of physiological hemocoagulation process [87, 103]. Fibrinogen for the preparation of fibrin could be isolated in reasonable quantities from the patient's own blood, that is used in autologous form [103]. Fibrin either enveloped individual fibers or formed additional nanofibrous network on the synthetic polymeric nanofibrous membranes. Fibrin was gradually degraded by the adhering and growing cells and was replaced by their own extracellular matrix, which was manifested by increased production of collagen in these cells. The cell growth and collagen production were further enhanced by the presence of ascorbic acid in the culture medium [87].

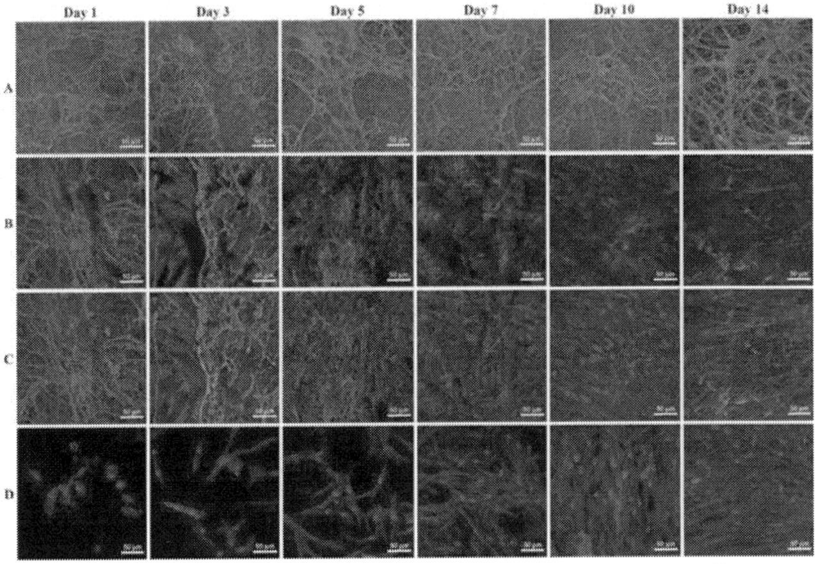

Figure 3. Morphology of fibrin coatings (green immunofluorescence) on nanofibrous PLA membranes in six time intervals incubated without cells at 37°C, 5% CO_2 (A), or incubated with human dermal fibroblasts (B—only fibrin, C—fibrin with cells). (D) Cells on non-modified PLA membranes. The cells were stained with phalloidin-TRITC and Hoechst #33258. Leica TCS SPE DM2500 confocal microscope, obj. 40×/1.15 NA oil, scale bar = 50 μm.

In another set of experiments, we studied the adhesion and growth of human keratinocytes on nanofibrous membranes made of poly-ε-caprolactone (PCL) and its copolymer with PLA (PLA/PCL, ratio 70:30). PCL and PLA/PCL copolymers have been experimentally used for vascular tissue engineering, particularly for replacement of small caliber blood vessels [31, 32], neural tissue engineering, specifically for generation of conductive sheaths for neurite outgrowth [104], for substituting the *dura mater* [105], and also for bone tissue engineering in order to mimic hemi-osteons and to control the spatial organization of osteoblasts [106]. These applications of PCL and PLA/PCL were enabled, among others, by suitable mechanical properties of these polymers, particularly in case of PLA/PCL. PLA/PCL also proved as suitable carriers for controlled delivery of drugs, for example antibiotics [107]. However, the potential of PCL and PLA/PCL in skin tissue engineering has not yet been fully explored. Our preliminary results with electrospun aliphatic polyesters, namely PCL and a PLA/PCL copolymer (kindly provided by the Technical University of Liberec, Faculty of Textile Engineering, Liberec, Czech Republic), showed that PCL and particularly PLA/PCL nanofibrous membranes would be suitable scaffolds for the adhesion and growth of skin cells. On PCL, human HaCaT keratinocytes were able to form large islands on day 7 after seeding (cell seeding density of 15,000 cells/cm^2), and on the PLA/PCL copolymer (ratio 70:30), even a confluent layer similar to that achieved on standard cell culture polystyrene dishes (**Figure 4**). The activity of mitochondrial enzymes, measured by the WST-1 test, showed a similar trend (**Figure 5**). The beneficial effect of the PLA/PCL copolymer was probably due to its higher hydrophilicity, and also to a greater thickness of the PLA/PCL fibers (diameter 1000 nm compared to 500 nm in PCL fibers), which might provide a better adhesion and growth support for cells. The interaction of skin cells with the PLA/PCL nanofibrous scaffolds can be further improved by loading these scaffolds with ascorbic acid [86], vitamin E, and curcumin [88].

Figure 4. Human HaCaT keratinocytes on day 7 after seeding on nanofibrous membranes made (A) of poly-ε-caprolactone (PCL), (B) of a copolymer of PCL and polylactide (PLA/PCL) and (C) on standard cell culture polystyrene dishes (PS). Cells stained with phalloidin conjugated with TRITC (red fluorescence), and the cell nuclei counterstained with DAPI (blue fluorescence). Olympus IX 71 microscope, DP 70 digital camera, scale bar = 200 μm.

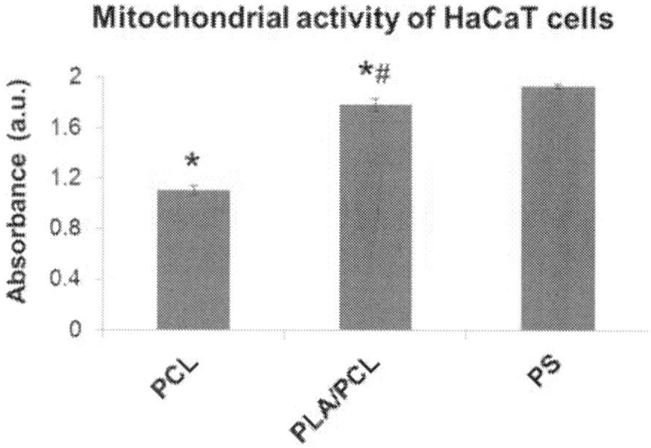

Figure 5. Activity of mitochondrial enzymes, measured by the WST-1 test in human HaCaT keratinocytes on day 7 after seeding on nanofibrous meshes made of poly-ε-caprolactone (PCL) and its copolymer with PLA (PLA/PCL), and on standard cell culsture polystyrene dishes (PS). Mean ± SEM (standard error of mean) from nine measurements for each experimental group. ANOVA, Student–Newman–Keuls method. Statistical significance: *#: $p \leq 0.05$ in comparison with PS and PCL, respectively.

Other promising nanofibrous scaffolds for skin tissue engineering are made of cellulose-based materials, which have achieved a remarkably wide range of applications in clinical practice. These materials serve as wound dressings, carriers for drug delivery, preparations for treatment of ophthalmological disorders, membranes for prevention of postoperative adhesions, meshes for hernia repair, materials for hemostasis, membranes for hemodialysis, and also as materials for plastic, reconstructive, and aesthetic surgery (for a review, see [108–110]). For tissue engineering, including skin tissue engineering, cellulose is promising due to its relatively good mechanical properties, low immunogenic properties, high biocompatibility, and water-holding ability [111, 112] (for a review, see [109, 110]). However, in its pure natural form, cellulose is nondegradable in the mammalian organism, while in tissue engineering, particularly that of skin, degradability of material scaffolds is necessary in order to achieve perfect skin regeneration without scar formation. Cellulose can be specifically degraded by cellulase, which hydrolyzes the 1,4-D-glycosidic linkage in the cellulose molecules and is produced by fungi, bacteria, and protozoans (for a review, see [113]). Thus, in preparation of scaffolds for tissue engineering, these enzymes have been incorporated in bacterial cellulose sheets [114]. Other approaches how to achieve and control the cellulose degradability were its esterification, that is formation of carboxymethylcellulose, which was then susceptible to degradation by esterases [115], acetylation, that is the formation of cellulose acetate, and its further mixing with another polysaccharide, pullulan, in various ratios [116]. Cellulose is also susceptible to hydrolysis by acids and, to a lesser extent, by alkalis (for a review, see [113]). Our earlier study showed that the degradation rate of cellulose can also be adjusted by percentage (wt.%) of COOH groups introduced into the cellulose molecules, but the following degradation of cellulose was accompanied by the release of glucuronic acid into the culture medium, which considerably lowered its pH and hampered the cell growth even at relatively low concentrations of COOH groups in the cellulose molecules (about 6 wt.% [108]).

Our recent experiments revealed that nanofibrous scaffolds made of cellulose acetate, acted as suitable growth support for human dermal fibroblasts *in vitro*. The cellulose acetate was purchased from Sigma–Aldrich (Cat. No. 180955), and the nanofibrous scaffolds were fabricated in Nanopharma Joint-Stock Co., Prague, Czech Republic. Neonatal human dermal fibroblasts (Lonza, Basel, Switzerland) were seeded on the scaffolds at the density of 15,000 cells/cm^2 and cultured in a Dulbecco's modified Eagle's medium (Sigma–Aldrich, Cat. No. D5648) supplemented with 10% of fetal bovine serum (FBS; Sebak GmbH, Aidenbach, Germany) and 40 μg/mL of gentamicin (LEK, Ljubljana, Slovenia). The number of the initially adhering cells on day 1 after seeding was lower on the nanofibrous cellulose scaffolds than on standard cell culture polystyrene dishes, and this number also remained lower on day 3 after seeding. The cell spreading was also lower on the nanofibrous cellulose acetate scaffolds than on polystyrene dishes. However, on day 7 after seeding, the cell numbers on the nanofibrous scaffolds and polystyrene dished almost equaled and reached confluence (**Figure 6**). Corresponding results were also obtained by the WST-1 test measuring the activity of mitochondrial enzymes (**Figure 7**).

Similarly, in a recent study, micro- and nanofibrous scaffolds made of cellulose acetate provided an excellent growth support for dermal fibroblasts in vitro, promoting a higher cell adhesion and mitochondrial activity (measured by the MTT test) compared with the control scaffolds made of PCL [117]. Positive influence on the adhesion and metabolic activity of human dermal fibroblasts were also obtained in composite nanofibrous scaffolds made of PCL and with cellulose acetate, cellulose acetate and pullulan [116], and particularly of electrospun cellulose acetate and gelatin in a ratio of 25:75 [118].

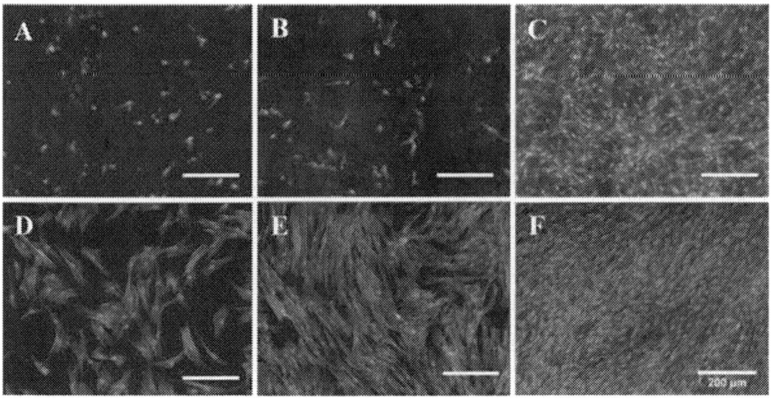

Figure 6. Human dermal fibroblasts on day 1 (A, D), day 3 (B, E), and day 7 (C, F) after seeding on nanofibrous membranes made of cellulose acetate (A–C) and on standard cell culture polystyrene dishes (D–F). Cells stained with phalloidin conjugated with TRITC (red fluorescence), and the cell nuclei counterstained with DAPI (blue fluorescence). Olympus IX 71 microscope, DP 70 digital camera, scale bar = 200 μm.

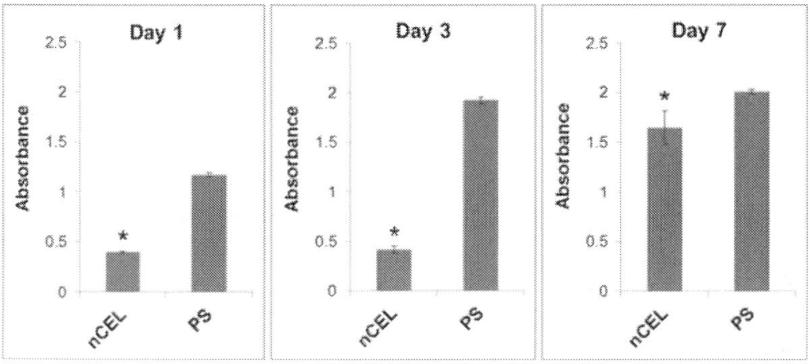

Figure 7. Activity of mitochondrial enzymes, measured by the WST-1 test in human dermal fibroblasts on days 1, 3, and 7 after seeding on nanofibrous membranes made of cellulose acetate (nCEL) and on standard cell culture polystyrene dishes (PS). Mean ± SEM (standard error of mean) from nine measurements for each experimental group and time ANOVA, Student–Newman–Keuls method. Statistical significance: *: $p \leq 0.05$ in comparison with PS.

3. Nanofibers in bone tissue engineering

Among a wide range of nanofibrous scaffolds used for bone tissue engineering, our studies concentrated on scaffolds reinforced with hydroxyapatite nanoparticles, diamond nanoparticles, and nanofibrous scaffolds created by the deposition of nanodiamond on SiO_2 nanofibers.

Nanofibrous PLA-hydroxyapatite (HAp) composites were created by addition of hydroxyapatite nanoparticles in concentrations of 5 wt.% and 15 wt.% to a PLA matrix before electrospinning. The addition of nanoparticles improved mechanical properties of the scaffolds by suppressing their creep behavior in their dry state. Addition of HAp nanoparticles also increased the proliferation of human osteoblast-like MG 63 cells, and particularly their osteogenic differentiation, manifested by production of osteocalcin, an extracellular matrix glycoprotein binding calcium [119].

Diamond nanoparticles were added either into PLGA or PLA matrix before electrospinning. PLGA nanofibers were enriched with 23 wt.% of diamond nanoparticles, and PLA nanofibers were enriched with several concentrations of diamond nanoparticles, ranging from 0.44 to 12.28 wt.%. To the best of our knowledge, we were the first laboratory creating nanodiamond-loaded polymeric nanofibrous scaffolds for potential bone tissue engineering. Earlier, nanofibrous polymer–nanodiamond composites were prepared only for technical applications, for example protection of various surfaces against scratch and potential damage by the ultraviolet light irradiation [120].

Our PLGA-nanodiamond nanofibrous scaffolds supported the adhesion and growth of human osteoblast-like MG 63 cells to a similar degree as the pure PLGA nanofibrous scaffolds [68] but accelerated the growth of human bone marrow mesenchymal stem cells [61] (**Figure 8**).

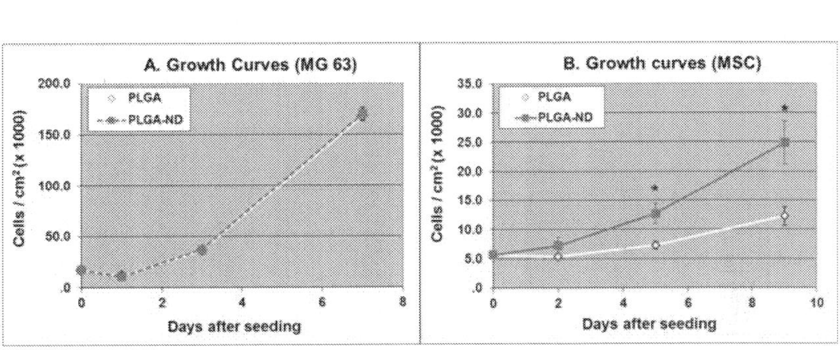

Figure 8. Growth curves of human osteoblast-like MG 63 cells (A) and human bone marrow mesenchymal stem cells (MSC, B) in cultures on pure poly(lactide-*co*-glycolide) (PLGA) scaffolds and scaffolds loaded with 23 wt.% of diamond nanoparticles (PLGA-ND). Mean ± SEM (standard error of mean) from 8 to 28 measurements for each experimental group and time interval. ANOVA, Student–Newman–Keuls method. Statistical significance: $^{*}p \leq 0.05$ in comparison with pure PLGA membranes.

However, when nanodiamond nanoparticles were incorporated into PLA nanofibrous scaffolds, they had rather negative effects on human osteoblast-like MG 63 and Saos-2 cells. The number and mitochondrial activity of cells growing of these scaffolds (**Figure 9**), as well as their expression of alkaline phosphatase and osteocalcin on the mRNA and protein levels decreased with increasing diamond particle concentration. This discrepancy was probably due to the different origin and different physicochemical properties of the diamond nanoparticles used for addition into PLGA and PLA nanofibers. For PLGA nanofibers, diamond nanoparticles were prepared by a radio-frequency PACVD method, while the PLA nanofibers were loaded with detonation nanodiamonds with hydrophobic surface (purchased from the Nano Carbon Research Institute, Japan, under the product name NanoAmando) (for a review, see [121]).

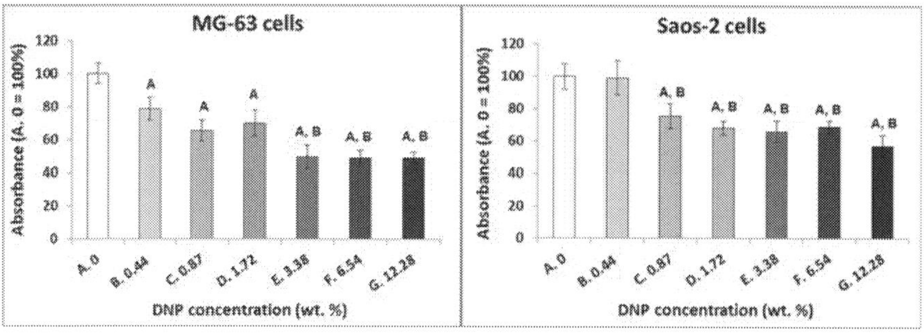

Figure 9. The mitochondrial activity of human osteoblast-like MG 63 and Saos-2 cells, measured by XTT test on day 3 after seeding on nanofibrous polylactide membranes loaded with 0–12.28 wt.% of diamond nanoparticles (DNP). Absorbances are given in % of values obtained from pure PLA membranes (sample A). Mean ± S.E.M. from 17 to 22 measurements for each experimental group and cell type. ANOVA, Student–Newman–KeulsStudent–Newman–Keuls method. Statistical significance: $^{A,\,B}p \leq 0.05$ in comparison with pure PLA membranes and membranes with the lowest DNP concentration, respectively.

Interesting results were obtained with a novel nanofibrous material, that is SiO_2 nanofibers prepared by electrospinning and then coated by a thin diamond film. SiO_2 nanofibers were purchased from the Technical University of Liberec, Czech Republic. Nanodiamond coating was performed by microwave plasma chemical vapor deposition [122]. Finally, both nanodiamond-coated and pure SiO_2 nanofibers were terminated by oxygen in order to enhance their attractiveness for the cell adhesion and growth. The nanofibers were seeded with human umbilical vein endothelial cells (HUVEC, passage 4) purchased from Lonza (Cat. No. C2517A) at the density of approx. 16,000 cells/cm^2, and the cells were cultured in endothelial growth medium (EGM-2, Lonza, Cat. No. CC-3162) for 1 and 4 days. The cells were then visualized using a Live/Dead Viability/Cytotoxicity assay kit (Life Technologies). The endothelial cells were chosen because they are an important cell type present in the bone, playing a key role in the scaffold vascularization. In addition, primary and low-passaged endothelial cells are a relatively demanding cell type sensitive to the physical and chemical properties of the material and its potential cytotoxicity. We found that the diamond coating on SiO_2 nanofibers markedly improved the growth of HUVEC cells. On day 1 after seeding, the number of viable cells was similar on both pure and diamond-coated SiO_2 nanofibers. However, on day 4, the cell number on pure SiO_2 remained similar as on day 1, while on diamond-coated SiO_2 nanofibers, it increased, and the cells formed islands (**Figure 10**).

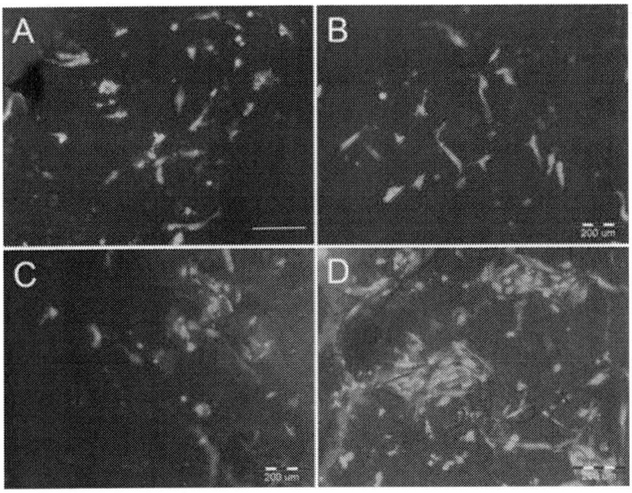

Figure 10. Human umbilical vein endothelial cells (HUVEC) cells grown on O-terminated SiO_2 nanofibers (A, B) and on O-terminated diamond-SiO_2 nanofibers on day 1 (A, C) and on day 4 (B, D) after seeding. The cells were stained with Live/Dead Viability/Cytotoxicity assay kit; living cells are stained in green and dead cells are stained in red. Olympus IX 71 microscope, IX71 digital camera, scale bar = 200 μm.

4. Conclusions

Nanofibrous scaffolds is one of the most promising materials for tissue engineering. At the experimental level, they have been used for construction or regeneration of almost all tissues in the human organism. Nanofibers are also applicable for the drug and gene delivery, gene silencing, biosensing, electrical stimulation of cells, and wound dressings. Nanofibers can be fabricated from a wide range of materials, mainly from natural and synthetic polymers, but also from ceramic and carbon materials, including SiO_2 and diamond. The main method recently used for fabrication of nanofibers is electrospinning. In order to enhance the attractiveness of nanofibers for the cell adhesion and growth, they can be loaded with various growth, angiogenic, and differentiation factors and/or functionalized with oligopeptidic ligands for the cell adhesion receptors. Other important modifications which improved the cell behavior on nanofibrous scaffolds were the activation of nanofibers by plasma treatment or coating of nanofibers with fibrin, as revealed by our earlier studies. For hard tissue engineering, the nanofibers can be reinforced with ceramic, metal-, or carbon- based nanoparticles, or by biomineralization. In our earlier studies, addition of hydroxyapatite to synthetic polymeric nanofibers increased the growth and osteogenic differentiation of human osteoblast-like cells. However, the addition of diamond nanoparticles to these nanofibers had controversial effects, which depended on the preparation method and physicochemical properties of the nanoparticles. Diamond nanoparticles prepared by PACVD method had stimulatory effects on the cell adhesion and growth, while the effects of diamond nanoparticles prepared by detonation synthesis were rather negative. A novel material promising for tissue engineering is the diamond-coated SiO_2 nanofiber, recently developed and tested by our group.

Acknowledgements

Our studies included in this review were supported by the Agency for Healthcare Research, Ministry of Health of the Czech Republic (grant no. 15-33018A), the Grant Agency of the Czech Republic (grant no. 14-04790S), and the Technology Grant Agency of the of the Czech Republic (grant no. TA04010065).

Author details

Lucie Bacakova[1*], Marketa Bacakova[1], Julia Pajorova[1], Radmila Kudlackova[1], Lubica Stankova[1], Elena Filova[1], Jana Musilkova[1], Stepan Potocky[2] and Alexander Kromka[2]

*Address all correspondence to: lucie.bacakova@fgu.cas.cz

1 Institute of Physiology, Czech Academy of Sciences, Videnska, Prague, Czech Republic

2 Institute of Physics, Czech Academy of Sciences, Cukrovarnicka, Prague, Czech Republic

References

[1] Sampath M, Lakra R, Korrapati P, Sengottuvelan B. Curcumin loaded poly (lactic-co-glycolic) acid nanofiber for the treatment of carcinoma. *Colloids and Surfaces B: Biointerfaces* 2014; 117: 128–134. doi:10.1016/j.colsurfb.2014.02.020.

[2] Xue J, He M, Niu Y, Liu H, Crawford A, Coates P, Chen D, Shi R, Zhang L. Preparation and in vivo efficient anti-infection property of GTR/GBR implant made by metronidazole loaded electrospun polycaprolactone nanofiber membrane. *International Journal of Pharmaceutics* 2014; 475(1–2): 566–577. doi:10.1016/j.ijpharm.2014.09.026.

[3] Meier C, Weil T, Kirchhoff F, Münch J. Peptide nanofibrils as enhancers of retroviral gene transfer. *Wiley Interdisciplinary Reviews: Nanomedicine and Nanobiotechnology* 2014; 6(5): 438–451. doi:10.1002/wnan.1275.

[4] Monteiro N, Ribeiro D, Martins A, Faria S, Fonseca NA, Moreira JN, Reis RL, Neves NM. Instructive Nanofibrous Scaffold Comprising Runt-Related Transcription Factor

2 Gene Delivery for Bone Tissue Engineering. *ACS Nano* 2014; 8(8): 8082–8094. doi: 10.1021/nn5021049.

[5] Goonoo N, Jeetah R, Bhaw-Luximon A, Jhurry D. Polydioxanone-based bio-materials for tissue engineering and drug/gene delivery applications. *European Journal of Pharmaceutics and Biopharmaceutics* 2015; 97: 371–391. doi:10.1016/j.ejpb.2015.05.024.

[6] Diao HJ, Low WC, Lu QR, Chew SY. Topographical effects on fiber-mediated microRNA delivery to control oligodendroglial precursor cells development. *Biomaterials* 2015; 70: 105–114. doi:10.1016/j.biomaterials.2015.08.029.

[7] Khadka DB, Haynie DT. Protein- and peptide-based electrospun nanofibers in medical biomaterials. *Nanomedicine: Nanotechnology, Biology and Medicine* 2012; 8(8): 1242–1262. doi:10.1016/j.nano.2012.02.013.

[8] Dallas P, Georgakilas V. Interfacial polymerization of conductive polymers: Generation of polymeric nanostructures in a 2-D space. *Advances in Colloid and Interface Science* 2015; 224: 46–61. doi:10.1016/j.cis.2015.07.008.

[9] Arslan A, Şimşek M, Aldemir SD, Kazaro140ğlu NM, Gümüşderelioğlu M. Honey-based PET or PET/chitosan fibrous wound dressings: effect of honey on electrospinning process. *Journal of Biomaterials Science, Polymer Edition* 2014; 25(10): 999–1012. doi: 10.1080/09205063.2014.918455.

[10] Unnithan AR, Gnanasekaran G, Sathishkumar Y, Lee YS, Kim CS. Electrospun antibacterial polyurethane–cellulose acetate–zein composite mats for wound dressing. *Carbohydrate Polymers* 2014; 102: 884–892. doi:10.1016/j.carbpol.2013.10.070.

[11] Zhong, Teo WE, Zhu X, Beuerman R, Ramakrishna S, Yung LYL. Formation of collagen–glycosaminoglycan blended nanofibrous scaffolds and their biological properties. *Biomacromolecules* 2005; 6(6): 2998–3004. doi:10.1021/bm050318p.

[12] Cao H, Chen MM, Liu Y, Liu YY, Huang YQ, Wang JH, Chen JD, Zhang QQ. Fish collagen-based scaffold containing PLGA microspheres for controlled growth factor delivery in skin tissue engineering. *Colloids and Surfaces B: Biointerfaces* 2015; 136: 1098–1106. doi:10.1016/j.colsurfb.2015.10.022.

[13] Sell SA, McClure MJ, Garg K, Wolfe PS, Bowlin GL. Electrospinning of collagen/biopolymers for regenerative medicine and cardiovascular tissue engineering. *Advanced Drug Delivery Reviews* 2009; 61(12): 1007–1019. doi:10.1016/j.addr.2009.07.012.

[14] Zeng L, Teng W, Jiang L, Cappello J, Wu X. Ordering recombinant silk-elastin-like nanofibers on the microscale. *Applied Physics Letters* 2014; 104(3): 033702. doi: 10.1063/1.4863077.

[15] Min BM, Lee G, Kim SH, Nam YS, Lee TS, Park WH. Electrospinning of silk fibroin nanofibers and its effect on the adhesion and spreading of normal human keratinocytes and fibroblasts in vitro. *Biomaterials* 2004; 25(7–8): 1289–1297. doi:10.1016/j.biomaterials.2003.08.045.

[16] Shao W, He J, Sang F, Ding B, Chen L, Cui S, Li K, Han Q, Tan W. Coaxial electrospun aligned tussah silk fibroin nanostructured fiber scaffolds embedded with hydroxyapatite–tussah silk fibroin nanoparticles for bone tissue engineering. *Materials Science and Engineering: C* 2016; 58: 342–351. doi:10.1016/j.msec.2015.08.046.

[17] Wang W, Itoh S, Konno K, Kikkawa T, Ichinose S, Sakai K, Ohkuma T, Watabe K. Effects of Schwann cell alignment along the oriented electrospun chitosan nanofibers on nerve regeneration. *Journal of Biomedical Materials Research Part A* 2009; 91A(4): 994–1005. doi:10.1002/jbm.a.32329.

[18] Shi Q, Li Y, Sun J, Zhang H, Chen L, Chen B, Yang H, Wang Z. The osteogenesis of bacterial cellulose scaffold loaded with bone morphogenetic protein-2. *Biomaterials* 2012; 33(28): 6644–6649. doi:10.1016/j.biomaterials.2012.05.071.

[19] Yao S, Wang X, Liu X, Wang R, Deng C, Cui F. Effects of ambient relative humidity and solvent properties on the electrospinning of pure hyaluronic acid nanofibers. *Journal of Nanoscience and Nanotechnology* 2013; 13(7): 4752–4758. doi:10.1166/jnn.2013.7197.

[20] Deng Y, Zhang X, Zhao Y, Liang S, Xu A, Gao X, Deng F, Fang J, Wei S. Peptide-decorated polyvinyl alcohol/hyaluronan nanofibers for human induced pluripotent stem cell culture. *Carbohydrate Polymers* 2014; 101: 36–39. doi:10.1016/j.carbpol.2013.09.030.

[21] Leszczak V, Place LW, Franz N, Popat KC, Kipper MJ. Nanostructured biomaterials from electrospun demineralized bone matrix: a survey of processing and crosslinking strategies. *ACS Applied Materials & Interfaces* 2014; 6(12): 9328–9337. doi:10.1021/am501700e.

[22] Ainslie KM, Bachelder EM, Borkar S, Zahr AS, Sen A, Badding JV, Pishko MV. Cell Adhesion on nanofibrous polytetrafluoroethylene (nPTFE). *Langmuir* 2007; 23(2): 747–754. doi:10.1021/la060948s.

[23] Hu Z jun, Li Z lun, Hu L yu, He W, Liu R ming, Qin Y sen, Wang S ming. The in vivo performance of small-caliber nanofibrous polyurethane vascular grafts. *BMC Cardiovascular Disorders* 2012; 12(1): 115. doi:10.1186/1471-2261-12-115.

[24] Dou Y, Wu C, Chang J. Preparation, mechanical property and cytocompatibility of poly(l-lactic acid)/calcium silicate nanocomposites with controllable distribution of calcium silicate nanowires. *Acta Biomaterialia* 2012; 8(11): 4139–4150. doi:10.1016/j.actbio.2012.07.009.

[25] Cai Q, Shi Y, Shan D, Jia W, Duan S, Deng X, Yang X. Osteogenic differentiation of MC3T3-E1 cells on poly(l-lactide)/Fe_3O_4 nanofibers with static magnetic field exposure. *Materials Science and Engineering: C* 2015; 55: 166–173. doi:10.1016/j.msec.2015.05.002.

[26] Xing ZC, Chae WP, Huh MW, Park LS, Park SY, Kwak G, Yoon KB, Kang IK. In vitro anti-bacterial and cytotoxic properties of silver-containing poly(l-lactide-co-glycolide)

nanofibrous scaffolds. *Journal of Nanoscience and Nanotechnology* 2011; 11(1): 61–65. doi: 10.1166/jnn.2011.3551.

[27] Zhang S, Prabhakaran MP, Qin X, Ramakrishna S. Poly-3-hydroxybutyrate-co-3-hydroxyvalerate containing scaffolds and their integration with osteoblasts as a model for bone tissue engineering. *Journal of Biomaterials Applications* 2015; 29(10): 1394–1406. doi:10.1177/0885328214568467.

[28] Song W, Markel DC, Wang S, Shi T, Mao G, Ren W. Electrospun polyvinyl alcohol–collagen–hydroxyapatite nanofibers: a biomimetic extracellular matrix for osteoblastic cells. *Nanotechnology* 2012; 23(11): 115101. doi:10.1088/0957-4484/23/11/115101.

[29] Manjumeena R, Elakkiya T, Duraibabu D, Feroze Ahamed A, Kalaichelvan P, Venkatesan R. "Green" biocompatible organic–inorganic hybrid electrospun nanofibers for potential biomedical applications. *Journal of Biomaterials Applications* 2015; 29(7): 1039–1055. doi:10.1177/0885328214550011.

[30] Vines JB, Lim DJ, Anderson JM, Jun HW. Hydroxyapatite nanoparticle reinforced peptide amphiphile nanomatrix enhances the osteogenic differentiation of mesenchymal stem cells by compositional ratios. *Acta Biomaterialia* 2012; 8(11): 4053–4063. doi: 10.1016/j.actbio.2012.07.024.

[31] Sankaran KK, Vasanthan KS, Krishnan UM, Sethuraman S. Development and evaluation of axially aligned nanofibres for blood vessel tissue engineering: small-diameter aligned nanofibrous vascular graft. *Journal of Tissue Engineering and Regenerative Medicine* 2014; 8(8): 640–651. doi:10.1002/term.1566.

[32] Wang Y, Hu J, Jiao J, Liu Z, Zhou Z, Zhao C, Chang LJ, Chen YE, Ma PX, Yang B. Engineering vascular tissue with functional smooth muscle cells derived from human iPS cells and nanofibrous scaffolds. *Biomaterials* 2014; 35(32): 8960–8969. doi:10.1016/j.biomaterials.2014.07.011.

[33] Hussain A, Collins G, Yip D, Cho CH. Functional 3-D cardiac co-culture model using bioactive chitosan nanofiber scaffolds. *Biotechnology and Bioengineering* 2013; 110(2): 637–647. doi:10.1002/bit.24727.

[34] Khan M, Xu Y, Hua S, Johnson J, Belevych A, Janssen PML, Gyorke S, Guan J, Angelos MG. Evaluation of changes in morphology and function of human induced pluripotent stem cell derived cardiomyocytes (HiPSC-CMs) cultured on an aligned-nanofiber cardiac patch. *PLOS ONE* 2015; 10(5): e0126338. doi:10.1371/journal.pone.0126338.

[35] Jahnavi S, Kumary TV, Bhuvaneshwar GS, Natarajan TS, Verma RS. Engineering of a polymer layered bio-hybrid heart valve scaffold. *Materials Science and Engineering: C* 2015; 51: 263–273. doi:10.1016/j.msec.2015.03.009.

[36] Punnoose AM. Electrospun Type 1 collagen matrices using a novel benign solvent for cardiac tissue engineering. *Journal of Cellular Physiology* 2016; 231(3): 744. doi:10.1002/jcp.25115.

[37] Jana S, Leung M, Chang J, Zhang M. Effect of nano- and micro-scale topological features on alignment of muscle cells and commitment of myogenic differentiation. *Biofabrication* 2014; 6(3): 035012. doi:10.1088/1758-5082/6/3/035012.

[38] Ostrovidov S, Shi X, Zhang L, Liang X, Kim SB, Fujie T, Ramalingam M, Chen M, Nakajima K, Al-Hazmi F, Bae H, Memic A, Khademhosseini A. Myotube formation on gelatin nanofibers—multi-walled carbon nanotubes hybrid scaffolds. *Biomaterials* 2014; 35(24): 6268–6277. doi:10.1016/j.biomaterials.2014.04.021.

[39] Bradshaw M, Ho D, Fear MW, Gelain F, Wood FM, Iyer KS. Designer self-assembling hydrogel scaffolds can impact skin cell proliferation and migration. *Scientific Reports* 2014; 4: 6903. doi:10.1038/srep06903.

[40] Wang Z, Qian Y, Li L, Pan L, Njunge LW, Dong L, Yang L. Evaluation of emulsion electrospun polycaprolactone/hyaluronan/epidermal growth factor nanofibrous scaffolds for wound healing. *Journal of Biomaterials Applications* 2016; 30(6): 686–698. doi:10.1177/0885328215586907.

[41] Zhou T, Wang N, Xue Y, Ding T, Liu X, Mo X, Sun J. Development of biomimetic tilapia collagen nanofibers for skin regeneration through inducing keratinocytes differentiation and collagen synthesis of dermal fibroblasts. *ACS Applied Materials & Interfaces* 2015; 7(5): 3253–3262. doi:10.1021/am507990m.

[42] Sahoo S, Ang LT, Cho-Hong Goh J, Toh SL. Bioactive nanofibers for fibroblastic differentiation of mesenchymal precursor cells for ligament/tendon tissue engineering applications. *Differentiation* 2010; 79(2): 102–110. doi:10.1016/j.diff.2009.11.001.

[43] Naghashzargar E, Farè S, Catto V, Bertoldi S, Semnani D, Karbasi S, Tanzi MC. Nano/micro hybrid scaffold of PCL or P3HB nanofibers combined with silk fibroin for tendon and ligament tissue engineering. *Journal of Applied Biomaterials & Functional Materials* 2015; 13(2): 0–0. doi:10.5301/jabfm.5000216.

[44] Boomer L, Liu Y, Mahler N, Johnson J, Zak K, Nelson T, Lannutti J, Besner GE. Scaffolding for challenging environments: materials selection for tissue engineered intestine: scaffolding for challenging environments. *Journal of Biomedical Materials Research Part A* 2014; 102(11): 3795–3802. doi:10.1002/jbm.a.35047.

[45] Kobayashi M, Lei NY, Wang Q, Wu BM, Dunn JCY. Orthogonally oriented scaffolds with aligned fibers for engineering intestinal smooth muscle. *Biomaterials* 2015; 61: 75–84. doi:10.1016/j.biomaterials.2015.05.023.

[46] Jang YS, Jang CH, Cho YB, Kim M, Kim GH. Tracheal regeneration using polycaprolactone/collagen-nanofiber coated with umbilical cord serum after partial resection. *International Journal of Pediatric Otorhinolaryngology* 2014; 78(12): 2237–2243. doi:10.1016/j.ijporl.2014.10.022.

[47] Mahoney C, Conklin D, Waterman J, Sankar J, Bhattarai N. Electrospun nanofibers of poly(ε-caprolactone)/depolymerized chitosan for respiratory tissue engineering

applications. *Journal of Biomaterials Science, Polymer Edition* 2016; 27(7): 611–625. doi: 10.1080/09205063.2016.1144454.

[48] Shakhssalim N, Rasouli J, Moghadasali R, Aghdas FS, Naji M, Soleimani M. Bladder smooth muscle cells interaction and proliferation on PCL/PLLA electrospun nanofibrous scaffold. *The International Journal of Artificial Organs* 2013; 36(2): 113–120. doi: 10.5301/ijao.5000175.

[49] Wei G, Li C, Fu Q, Xu Y, Li H. Preparation of PCL/silk fibroin/collagen electrospun fiber for urethral reconstruction. *International Urology and Nephrology* 2015; 47(1): 95–99. doi: 10.1007/s11255-014-0854-3.

[50] Mareková D, Lesný P, Jendelová P, Michálek J, Kostecká P, Přádný M, Martinová L, Pantoflíček T, Ryska M, Syková E. Hepatocyte growth on polycapronolactone and 2-hydroxyethylmethacrylate nanofiber sheets enhanced by bone marrow-derived mesenchymal stromal cells. *Hepato-Gastroenterology* 2013; 60(125): 1156–1163. doi: 10.5754/hge11948.

[51] Bishi DK, Mathapati S, Venugopal JR, Guhathakurta S, Cherian KM, Verma RS, Ramakrishna S. A patient-inspired ex vivo liver tissue engineering approach with autologous mesenchymal stem cells and hepatogenic serum. *Advanced Healthcare Materials* 2016: n/a–n/a. doi:10.1002/adhm.201500897.

[52] Chow LW, Wang L jia, Kaufman DB, Stupp SI. Self-assembling nanostructures to deliver angiogenic factors to pancreatic islets. *Biomaterials* 2010; 31(24): 6154–6161. doi: 10.1016/j.biomaterials.2010.04.002.

[53] Uzunalli G, Tumtas Y, Delibasi T, Yasa O, Mercan S, Guler MO, Tekinay AB. Improving pancreatic islet in vitro functionality and transplantation efficiency by using heparin mimetic peptide nanofiber gels. *Acta Biomaterialia* 2015; 22: 8–18. doi:10.1016/j.actbio.2015.04.032.

[54] Fon D, Zhou K, Ercole F, Fehr F, Marchesan S, Minter MR, Crack PJ, Finkelstein DI, Forsythe JS. Nanofibrous scaffolds releasing a small molecule BDNF-mimetic for the re-direction of endogenous neuroblast migration in the brain. *Biomaterials* 2014; 35(9): 2692–2712. doi:10.1016/j.biomaterials.2013.12.016.

[55] Sudwilai T, Ng JJ, Boonkrai C, Israsena N, Chuangchote S, Supaphol P. Polypyrrole-coated electrospun poly(lactic acid) fibrous scaffold: effects of coating on electrical conductivity and neural cell growth. *Journal of Biomaterials Science, Polymer Edition* 2014; 25(12): 1240–1252. doi:10.1080/09205063.2014.926578.

[56] Rafalowska J, Sulejczak D, Chrapusta SJ, Gadamski R, Taraszewska A, Nakielski P, Kowalczyk T, Dziewulska D. Original article Non-woven nanofiber mats—a new perspective for experimental studies of the central nervous system? *Folia Neuropathologica* 2014; 4: 407–416. doi:10.5114/fn.2014.47841.

[57] Wang J, Zheng J, Zheng Q, Wu Y, Wu B, Huang S, Fang W, Guo X. FGL-functionalized self-assembling nanofiber hydrogel as a scaffold for spinal cord-derived neural

stem cells. *Materials Science and Engineering: C* 2015; 46: 140–147. doi:10.1016/j.msec.2014.10.019.

[58] Yan L, Zhao B, Liu X, Li X, Zeng C, Shi H, Xu X, Lin T, Dai L, Liu Y. Aligned nanofibers from polypyrrole/graphene as electrodes for regeneration of optic nerve via electrical stimulation. *ACS Applied Materials & Interfaces* 2016; 8(11): 6834–6840. doi:10.1021/acsami.5b12843.

[59] Popelka Š, Studenovská H, Abelová L, Ardan T, Studený P, Straňák Z, Klíma J, Dvořánková B, Kotek J, Hodan J, Rypáček F. A frame-supported ultrathin electrospun polymer membrane for transplantation of retinal pigment epithelial cells. *Biomedical Materials* 2015; 10(4): 045022. doi:10.1088/1748-6041/10/4/045022.

[60] Xia H, Chen Q, Fang Y, Liu D, Zhong D, Wu H, Xia Y, Yan Y, Tang W, Sun X. Directed neurite growth of rat dorsal root ganglion neurons and increased colocalization with Schwann cells on aligned poly(methyl methacrylate) electrospun nanofibers. *Brain Research* 2014; 1565: 18–27. doi:10.1016/j.brainres.2014.04.002.

[61] Brady MA, Renzing A, Douglas TEL, Liu Q, Wille S, Parizek M, Bacakova L, Kromka A, Jarosova M, Godier G, Warnke PH. Development of composite poly(lactide-coglycolide)-nanodiamond scaffolds for bone cell growth. *Journal of Nanoscience and Nanotechnology* 2015; 15(2): 1060–1069. doi:10.1166/jnn.2015.9745.

[62] Duan S, Yang X, Mei F, Tang Y, Li X, Shi Y, Mao J, Zhang H, Cai Q. Enhanced osteogenic differentiation of mesenchymal stem cells on poly(L-lactide) nanofibrous scaffolds containing carbon nanomaterials: enhanced osteocompatibility of nanofibrous scaffolds. *Journal of Biomedical Materials Research Part A* 2015; 103(4): 1424–1435. doi:10.1002/jbm.a.35283.

[63] Zhang X, Meng S, Huang Y, Xu M, He Y, Lin H, Han J, Chai Y, Wei Y, Deng X. Electrospun gelatin/β-TCP composite nanofibers enhance osteogenic differentiation of BMSCs and *in vivo* bone formation by activating Ca^{2+}-sensing receptor signaling. *Stem Cells International* 2015; 2015: 1–13. doi:10.1155/2015/507154.

[64] Kung FC, Lin CC, Lai WFT. Osteogenesis of human adipose-derived stem cells on hydroxyapatite-mineralized poly(lactic acid) nanofiber sheets. *Materials Science and Engineering: C* 2014; 45: 578–588. doi:10.1016/j.msec.2014.10.005.

[65] Mertaniemi H, Escobedo-Lucea C, Sanz-Garcia A, Gandía C, Mäkitie A, Partanen J, Ikkala O, Yliperttula M. Human stem cell decorated nanocellulose threads for biomedical applications. *Biomaterials* 2016; 82: 208–220. doi:10.1016/j.biomaterials.2015.12.020.

[66] Erisken C. Viscoelastic and biomechanical properties of osteochondral tissue constructs generated from graded polycaprolactone and beta-tricalcium phosphate composites. *Journal of Biomechanical Engineering* 2010; 132(9): 091013. doi:10.1115/1.4001884.

[67] Münchow EA, Pankajakshan D, Albuquerque MTP, Kamocki K, Piva E, Gregory RL, Bottino MC. Synthesis and characterization of CaO-loaded electrospun matrices for

bone tissue engineering. *Clinical Oral Investigations* 2015. doi:10.1007/s00784-015-1671-5.

[68] Parizek M, Douglas TE, Novotna K, Kromka A, Brady MA, Renzing A, Voss E, Jarosova M, Palatinus L, Tesarek P, Ryparova P, Lisa V, dos Santos AM, Bacakova L. Nanofibrous poly(lactide-co-glycolide) membranes loaded with diamond nanoparticles as promising substrates for bone tissue engineering. *International Journal of Nanomedicine* 2012: 1931. doi:10.2147/IJN.S26665.

[69] Joshi MK, Pant HR, Tiwari AP, Maharjan B, Liao N, kim HJ, Park CH, Kim CS. Three-dimensional cellulose sponge: fabrication, characterization, biomimetic mineralization, and in vitro cell infiltration. *Carbohydrate Polymers* 2016; 136: 154–162. doi:10.1016/j.carbpol.2015.09.018.

[70] Yang M, Zhou G, Castano-Izquierdo H, Zhu Y, Mao C. Biomineralization of natural collagenous nanofibrous membranes and their potential use in bone tissue engineering. *Journal of Biomedical Nanotechnology* 2015; 11(3): 447–456. doi:10.1166/jbn.2015.2038.

[71] Çakmak S, Çakmak AS, Gümüşderelioğlu M. RGD-bearing peptide-amphiphile-hydroxyapatite nanocomposite bone scaffold: an in vitro study. *Biomedical Materials* 2013; 8(4): 045014. doi:10.1088/1748-6041/8/4/045014.

[72] Pasuri J, Holopainen J, Kokkonen H, Persson M, Kauppinen K, Lehenkari P, Santala E, Ritala M, Tuukkanen J. Osteoclasts in the interface with electrospun hydroxyapatite. *Colloids and Surfaces B: Biointerfaces* 2015; 135: 774–783. doi:10.1016/j.colsurfb.2015.08.045.

[73] Jia X, Tang T, Cheng D, Zhang C, Zhang R, Cai Q, Yang X. Micro-structural evolution and biomineralization behavior of carbon nanofiber/bioactive glass composites induced by precursor aging time. *Colloids and Surfaces B: Biointerfaces* 2015; 136: 585–593. doi:10.1016/j.colsurfb.2015.09.062.

[74] Liu Y, Sagi S, Chandrasekar R, Zhang L, Hedin NE, Fong H. Preparation and characterization of electrospun SiO_2 nanofibers. *Journal of Nanoscience and Nanotechnology* 2008; 8(3): 1528–1536.

[75] Ganesh VA, Dinachali SS, Raut HK, Walsh TM, Nair AS, Ramakrishna S. Electrospun SiO_2 nanofibers as a template to fabricate a robust and transparent superamphiphobic coating. *RSC Advances* 2013; 3(12): 3819. doi:10.1039/c3ra22968h.

[76] Potocký š., Ižák T, Rezek B, Tesárek P, Kromka A. Transformation of polymer composite nanofibers to diamond fibers and films by microwave plasma-enhanced CVD process. *Applied Surface Science* 2014; 312: 188–191. doi:10.1016/j.apsusc.2014.05.119.

[77] Ruffinatto S, Girard HA, Becher F, Arnault JC, Tromson D, Bergonzo P. Diamond porous membranes: a material toward analytical chemistry. *Diamond and Related Materials* 2015; 55: 123–130. doi:10.1016/j.diamond.2015.03.008.

[78] Ralchenko VG, Sovyk DN, Bolshakov AP, Homich AA, Vlasov II, Kurdyukov DA, Golubev VG, Zakhidov AA. Diamond direct and inverse opal matrices produced by

chemical vapor deposition. *Physics of the Solid State* 2011; 53(6): 1131–1134. doi:10.1134/S106378341106028X.

[79] Gao F, Nebel CE. Diamond-based supercapacitors: realization and properties. *ACS Applied Materials & Interfaces* 2015. doi:10.1021/acsami.5b07027.

[80] Kabiri M, Oraee-Yazdani S, Shafiee A, Hanaee-Ahvaz H, Dodel M, Vaseei M, Soleimani M. Neuroregenerative effects of olfactory ensheathing cells transplanted in a multi-layered conductive nanofibrous conduit in peripheral nerve repair in rats. *Journal of Biomedical Science* 2015; 22(1) . doi:10.1186/s12929-015-0144-0.

[81] Fraczek-Szczypta A. Carbon nanomaterials for nerve tissue stimulation and regeneration. *Materials Science and Engineering: C* 2014; 34: 35–49. doi:10.1016/j.msec.2013.09.038.

[82] Kharaziha M, Shin SR, Nikkhah M, Topkaya SN, Masoumi N, Annabi N, Dokmeci MR, Khademhosseini A. Tough and flexible CNT–polymeric hybrid scaffolds for engineering cardiac constructs. *Biomaterials* 2014; 35(26): 7346–7354. doi:10.1016/j.biomaterials.2014.05.014.

[83] Dolati F, Yu Y, Zhang Y, Jesus AMD, Sander EA, Ozbolat IT. In vitro evaluation of carbon-nanotube-reinforced bioprintable vascular conduits. *Nanotechnology* 2014; 25(14): 145101. doi:10.1088/0957-4484/25/14/145101.

[84] Norouzi M, Shabani I, Ahvaz HH, Soleimani M. PLGA/gelatin hybrid nanofibrous scaffolds encapsulating EGF for skin regeneration: PLGA/gelatin hybrid nanofibrous scaffolds. *Journal of Biomedical Materials Research Part A* 2015; 103(7): 2225–2235. doi:10.1002/jbm.a.35355.

[85] Lai HJ, Kuan CH, Wu HC, Tsai JC, Chen TM, Hsieh DJ, Wang TW. Tailored design of electrospun composite nanofibers with staged release of multiple angiogenic growth factors for chronic wound healing. *Acta Biomaterialia* 2014; 10(10): 4156–4166. doi:10.1016/j.actbio.2014.05.001.

[86] Sridhar S, Venugopal JR, Ramakrishna S. Improved regeneration potential of fibroblasts using ascorbic acid-blended nanofibrous scaffolds: regeneration potential of ascorbic acid-blended nanofibrous scaffolds. *Journal of Biomedical Materials Research Part A* 2015; 103(11): 3431–3440. doi:10.1002/jbm.a.35486.

[87] Bacakova M, Musilkova J, Riedel T, Stranska D, Brynda E, Bacakova L, Zaloudkova M. The potential applications of fibrin-coated electrospun polylactide nanofibers in skin tissue engineering. *International Journal of Nanomedicine* 2016: 771. doi:10.2147/IJN.S99317.

[88] Srinivasan DK, Gandhimathi C, Venugopal JR, Bhaarathy V, Ramakrishna S. Biocomposite nanofibrous strategies for the controlled release of biomolecules for skin tissue regeneration. *International Journal of Nanomedicine* 2014: 4709. doi:10.2147/IJN.S65335.

[89] Li J, Fu R, Li L, Yang G, Ding S, Zhong Z, Zhou S. Co-delivery of dexamethasone and green tea polyphenols using electrospun ultrafine fibers for effective treatment of keloid. *Pharmaceutical Research* 2014; 31(7): 1632–1643. doi:10.1007/s11095-013-1266-2.

[90] Kim JI, Pant HR, Sim HJ, Lee KM, Kim CS. Electrospun propolis/polyurethane composite nanofibers for biomedical applications. *Materials Science and Engineering: C* 2014; 44: 52–57. doi:10.1016/j.msec.2014.07.062.

[91] Park SJ, Lee BK, Na MH, Kim DS. Melt-spun shaped fibers with enhanced surface effects: fiber fabrication, characterization and application to woven scaffolds. *Acta Biomaterialia* 2013; 9(8): 7719–7726. doi:10.1016/j.actbio.2013.05.001.

[92] Flaibani M, Elvassore N. Gas anti-solvent precipitation assisted salt leaching for generation of micro- and nano-porous wall in bio-polymeric 3D scaffolds. *Materials Science and Engineering: C* 2012; 32(6): 1632–1639. doi:10.1016/j.msec.2012.04.054.

[93] Pokorny P, Kostakova E, Sanetrnik F, Mikes P, Chvojka J, Kalous T, Bilek M, Pejchar K, Valtera J, Lukas D. Effective AC needleless and collectorless electrospinning for yarn production. *Phys Chem Chem Phys* 2014; 16(48): 26816–26822. doi:10.1039/C4CP04346D.

[94] Orgill D, Blanco C, editors. *Biomaterials for treating skin loss*. Boca Raton, FL: CRC Press; 2009.

[95] Wojtowicz AM, Oliveira S, Carlson MW, Zawadzka A, Rousseau CF, Baksh D. The importance of both fibroblasts and keratinocytes in a bilayered living cellular construct used in wound healing: two cell type importance in wound healing. *Wound Repair and Regeneration* 2014; 22(2): 246–255. doi:10.1111/wrr.12154.

[96] Ng KW, Tham W, Lim TC, Werner Hutmacher D. Assimilating cell sheets and hybrid scaffolds for dermal tissue engineering. *Journal of Biomedical Materials Research Part A* 2005; 75A(2): 425–438. doi:10.1002/jbm.a.30454.

[97] van Dorp AGM, Verhoeven MCH, Koerten HK, van Blitterswijk CA, Ponec M. Bilayered biodegradable poly(ethylene glycol)/poly(butylene terephthalate) copolymer (Polyactive?) as substrate for human fibroblasts and keratinocytes. *Journal of Biomedical Materials Research* 1999; 47(3): 292–300. doi:10.1002/(SICI)1097-4636(19991205)47:3<292::AID-JBM2>3.0.CO;2-B.

[98] Idrus RH, Rameli MA bin P, Low KC, Law JX, Chua KH, Latiff MBA, Saim AB. Full-thickness skin wound healing using autologous keratinocytes and dermal fibroblasts with fibrin: bilayered versus single-layered substitute. *Advances in Skin & Wound Care* 2014; 27(4): 171–180. doi:10.1097/01.ASW.0000445199.26874.9d.

[99] Marston WA, Sabolinski ML, Parsons NB, Kirsner RS. Comparative effectiveness of a bilayered living cellular construct and a porcine collagen wound dressing in the treatment of venous leg ulcers: effectiveness of BLCC vs. SIS in VLU. *Wound Repair and Regeneration* 2014; 22(3): 334–340. doi:10.1111/wrr.12156.

[100] Monteiro IP, Gabriel D, Timko BP, Hashimoto M, Karajanagi S, Tong R, Marques AP, Reis RL, Kohane DS. A two-component pre-seeded dermal–epidermal scaffold. *Acta Biomaterialia* 2014; 10(12): 4928–4938. doi:10.1016/j.actbio.2014.08.029.

[101] Wendt H, Hillmer A, Reimers K, Kuhbier JW, Schäfer-Nolte F, Allmeling C, Kasper C, Vogt PM. Artificial skin—Culturing of different skin cell lines for generating an artificial skin substitute on cross-weaved spider silk fibres. *PLoS One* 2011; 6(7): e21833. doi:10.1371/journal.pone.0021833.

[102] Bacakova M, Lopot F, Hadraba D, Varga M, Zaloudkova M, Stranska D, Suchy T, Bacakova L. Effects of fiber density and plasma modification of nanofibrous membranes on the adhesion and growth of HaCaT keratinocytes. *Journal of Biomaterials Applications* 2015; 29(6): 837–853. doi:10.1177/0885328214546647.

[103] Filová E, Brynda E, Riedel T, Chlupáč J, Vandrovcová M, Švindrych Z, Lisá V, Houska M, Pirk J, Bačáková L. Improved adhesion and differentiation of endothelial cells on surface-attached fibrin structures containing extracellular matrix proteins: endothelial cells on surface-attached fibrin. *Journal of Biomedical Materials Research Part A* 2014; 102(3): 698–712. doi:10.1002/jbm.a.34733.

[104] Xie J, MacEwan MR, Willerth SM, Li X, Moran DW, Sakiyama-Elbert SE, Xia Y. Conductive core-sheath nanofibers and their potential application in neural tissue engineering. *Advanced Functional Materials* 2009; 19(14): 2312–2318. doi:10.1002/adfm.200801904.

[105] Wang Y fei, Guo H feng, Ying D jun. Multilayer scaffold of electrospun PLA-PCL-collagen nanofibers as a dural substitute: PLA-PCL-collagen as a dural substitute. *Journal of Biomedical Materials Research Part B: Applied Biomaterials* 2013; 101(8): 1359–1366. doi:10.1002/jbm.b.32953.

[106] Nedjari S, Eap S, Hébraud A, Wittmer CR, Benkirane-Jessel N, Schlatter G. Electrospun honeycomb as nests for controlled osteoblast spatial organization: electrospun honeycomb as nests for controlled. *Macromolecular Bioscience* 2014; 14(11): 1580–1589. doi:10.1002/mabi.201400226.

[107] Valarezo E, Tammaro L, Malagón O, González S, Armijos C, Vittoria V. Fabrication and characterization of poly(lactic acid)/poly(ε-caprolactone) blend electrospun fibers loaded with amoxicillin for tunable delivering. *Journal of Nanoscience and Nanotechnology* 2015; 15(6): 4706–4712. doi:10.1166/jnn.2015.9726.

[108] Novotna K, Havelka P, Sopuch T, Kolarova K, Vosmanska V, Lisa V, Svorcik V, Bacakova L. Cellulose-based materials as scaffolds for tissue engineering. *Cellulose* 2013; 20(5): 2263–2278. doi:10.1007/s10570-013-0006-4.

[109] Bačáková L, Novotná K, Pařízek M. Polysaccharides as cell carriers for tissue engineering: the use of cellulose in vascular wall reconstruction. *Physiological Research/ Academia Scientiarum Bohemoslovaca* 2014; 63(Suppl. 1): S29–S47.

[110] Bacakova L, Novotna K, Sopuch T, Havelka P. Cell interaction with cellulose-based scaffolds for tissue engineering—a review. In: Cellulose and Cellulose Derivatives: Synthesis, Modification, Nanostructure and Applications. Ibrahim H. Mondal (ed.), Nova Science Publishers, Inc., New York, USA, 2015, chapter 13, pp. 341–376, ISBN: 978-1-63483-571-8.

[111] Kwak MH, Kim JE, Go J, Koh EK, Song SH, Son HJ, Kim HS, Yun YH, Jung YJ, Hwang DY. Bacterial cellulose membrane produced by *Acetobacter* sp. A10 for burn wound dressing applications. *Carbohydrate Polymers* 2015; 122: 387–398. doi:10.1016/j.carbpol.2014.10.049.

[112] Bottan S, Robotti F, Jayathissa P, Hegglin A, Bahamonde N, Heredia-Guerrero JA, Bayer IS, Scarpellini A, Merker H, Lindenblatt N, Poulikakos D, Ferrari A. Surface-structured bacterial cellulose with guided assembly-based biolithography (GAB). *ACS Nano* 2015; 9(1): 206–219. doi:10.1021/nn5036125.

[113] Song SH, Kim JE, Lee YJ, Kwak MH, Sung GY, Kwon SH, Son HJ, Lee HS, Jung YJ, Hwang DY. Cellulose film regenerated from Styela clava tunics have biodegradability, toxicity and biocompatibility in the skin of SD rats. *Journal of Materials Science: Materials in Medicine* 2014; 25(6): 1519–1530. doi:10.1007/s10856-014-5182-8.

[114] Hu Y, Catchmark JM. Integration of cellulases into bacterial cellulose: toward bioabsorbable cellulose composites. *Journal of Biomedical Materials Research Part B: Applied Biomaterials* 2011; 97B(1): 114–123. doi:10.1002/jbm.b.31792.

[115] Lee SY, Bang S, Kim S, Jo SY, Kim BC, Hwang Y, Noh I. Synthesis and in vitro characterizations of porous carboxymethyl cellulose-poly(ethylene oxide) hydrogel film. *Biomaterials Research* 2015; 19(1) . doi:10.1186/s40824-015-0033-3.

[116] Atila D, Keskin D, Tezcaner A. Cellulose acetate based 3-dimensional electrospun scaffolds for skin tissue engineering applications. *Carbohydrate Polymers* 2015; 133: 251–261. doi:10.1016/j.carbpol.2015.06.109.

[117] Kim M, Kim G. 3D multi-layered fibrous cellulose structure using an electrohydrodynamic process for tissue engineering. *Journal of Colloid and Interface Science* 2015; 457: 180–187. doi:10.1016/j.jcis.2015.07.007.

[118] Vatankhah E, Prabhakaran MP, Jin G, Mobarakeh LG, Ramakrishna S. Development of nanofibrous cellulose acetate/gelatin skin substitutes for variety wound treatment applications. *Journal of Biomaterials Applications* 2014; 28(6): 909–921. doi: 10.1177/0885328213486527.

[119] Novotna K, Zajdlova M, Suchy T, Hadraba D, Lopot F, Zaloudkova M, *et al.* Polylactide nanofibers with hydroxyapatite as growth substrates for osteoblast-like cells. *Journal of Biomedical Materials Research Part A* 2014; 102(11): 3918–3930. doi:10.1002/jbm.a.35061.

[120] Behler KD, Stravato A, Mochalin V, Korneva G, Yushin G, Gogotsi Y. Nanodiamond-polymer composite fibers and coatings. *ACS Nano* 2009; 3(2): 363–369. doi:10.1021/nn800445z.

[121] Bacakova L, Kopova I, Stankova L, Liskova J, Vacik J, Lavrentiev V, Kromka A, Potocky S, Stranska D. Bone cells in cultures on nanocarbon-based materials for potential bone tissue engineering: a review: bone cells in cultures on nanocarbon-based materials. *Physica Status Solidi (a)* 2014; 211(12): 2688–2702. doi:10.1002/pssa.201431402.

[122] Potocky S, Kromka A, Potmesil J, Remes Z, Polackova Z, Vanecek M. Growth of nanocrystalline diamond films deposited by microwave plasma CVD system at low substrate temperatures. *Physica Status Solidi (a)* 2006; 203(12): 3011–3015. doi:10.1002/pssa.200671110.

Chapter 4

Tailoring of Architecture and Intrinsic Structure of Electrospun Nanofibers by Process Parameters for Tissue Engineering Applications

Dorota Kołbuk

Additional information is available at the end of the chapter

http://dx.doi.org/10.5772/64177

Abstract

Electrospinning process is commercially used to form nanofibers as scaffolds in tissue engineering. Similarities in morphology of electrospun nanofibers to the natural extracellular matrix, flexibility, and low cost of the process contribute to their use in regeneration of cartilage, ligaments/tendons, muscles, and bones. Required properties are tailored by the use of appropriate polymers: polyesters, their copolymers, blends with natural biopolymers such as gelatin, collagen, chitosan, or composites with nanoparticles.

In the case of one component fibers, factors strongly influencing the final diameter of the electrospinning jet include volumetric charge density, distance between the needle and the collector, needle diameter, and viscosity. A moderate effect is exerted by initial polymer concentration, solution density, electric potential, and solvent vapor pressure.

In the case of blend fibers, the w/w% ratio of mixed polymers is an additional parameter of the electrospinning process. Addition of gelatin, collagen, and/or chitosan influence the solution properties and, in consequence, fiber diameter, mechanical properties, wettability, chemical structure, crystallinity, etc. Cellular response depends on electrospun fibers' tailored morphology, chemical structure as well as mechanical properties.

Electrospinning process is one of the success stories in nanotechnologies during the last few years. Understanding of electrospinning process parameters enables tailoring of electrospun nanofibers morphology, internal structure, and properties to appropriate application. This opens up new possibilities in tissue engineering.

Keywords: electrospinning, tissue engineering, collagen, chitosan, gelatin, polyester, nanofibers, scaffold

1. Introduction

Natural and synthetic fibers have been used commonly in many applications. Silk, cotton, cellulose fibers have been used in textile industry and in regenerative medicine. Fibers and patches obtained by melt spinning, solution wet spinning, solution dry spinning, dry-jet wet spinning, gel spinning, phase separation spinning, reaction spinning are performed in diameter from nano to hundreds microns [1].

Electrospinning is one of the most common methods of nanofiber formation for the purposes of life science, protective clothing, filters, nanosensors, tissue engineering, drug delivery systems, and other applications. Synthetic and natural polymers, their blends, and composites with appropriate nanoparticles are used in the electrospinning process to form nano and submicron fibers with architecture and properties suitable for appropriate applications [2]. Fibers formed in the electrospinning process are used in tissue engineering, regeneration of cartilage, ligaments/tendons, muscles, and bones. Generally, the polymer and bioactivator used in the electrospinning process determine application in a suitable direction.

About 30 different parameters influence transformation of a polymer solution into electrospun nanofibers [3, 4]. These parameters include (a) the solution properties such as viscosity, conductivity, and surface tension; (b) processing parameters such as the applied electric potential at the capillary tip and at the collector, distance between the tip and the collecting screen, solution flow rate; and (c) ambient parameters such as solution temperature, and humidity [5]. Due to many parameters, the electrospinning process is the most used multipurpose technique of nanofibers and patches formation.

In this chapter, we concentrate on nanofibers formed for tissue engineering applications. Langer and Vacanti in 1993 [6] defined tissue engineering as an interdisciplinary field that applies the principles of engineering and life sciences toward the development of biological substitutes that restore, maintain, or improve function (of tissue) or a whole organ. Ideally, a fibrous patch should mimic the shape and functionality of the native extracellular matrix (ECM) [7]. Scaffolds formed in the electrospinning process show great potential to provide ideal microenvironments for growth and proliferation of various kinds of cells. Synthetic, natural polymers and their blends are used for electrospinning to combine advantages of both components. Electrospinning patches show various morphology, architecture, surface, and mechanical properties, which have strong influence on cell growth, spreading, activity, and functions [8].

It is widely described that the morphology of electrospun nanofibers strongly depends on the solution properties (determined by the type of polymer and solvent (for example, see [3, 4, 10]) and parameters of the process (applied voltage, flow rate, diameter of the needle (for example, see [11, 12]). Several studies describe the influence of the aforementioned parameters on the molecular structure. Crystallinity of poly(ε-caprolactone) (PCL) has been described to depend on the solvent type [13]. Investigations on poly(vinyl alcohol) (PVA) indicate an increase of the molecular orientation with an increase of the polymer molecular weight (Mn) [14]. Still, only few publications show influence of applied voltage on crystallinity of polyesters used in tissue engineering. Investigations on cellulose nanofibers (collected on a static plate) indicate an increase of crystallinity, with a local maximum value, as a function of applied voltage [15]. Qualitative analyses of crystallinity and molecular orientation of electrospun polyacrylonitrile (PAN) and poly(lactic acid) (PLLA) nanofibers indicate changes depending on polymer concentration, rotation speed of the collector (drum), and thermal treatment. Surface properties, such as stiffness, strongly depend on the molecular structure. Recent reports of in-vitro studies suggest a relationship between cellular response and molecular structure [16, 17].

Naturally occurring polymers, such as collagen (Col), Gelatin (Ge) and chitosan (Chit), implanted into humans do not cause foreign body response so they are successfully used in tissue engineering [18]. Major drawbacks are poor mechanical properties and variable physical properties due to different sources of the protein matrices. Taking these criteria into account, a hybrid of a natural polymer, such as Col, Ge or Chit, and e.g. PCL, represents a biocompatible and degradable polymer [19]. Bicomponent nanofibers combine the advantages of both types of materials—natural and synthetic (mechanical properties and reproducibility).

Proteins like Col and Ge in fibres form simulate Col fibres in ECM. Chit is a semi-crystalline polysaccharide which demonstrates high similarity to glycosaminoglycan also in ECM [20]. Due to this resemblance they are considered as the main or partial additive in electrospun nanofibers for tissues regeneration and antibacterial wound dressings.

This chapter sums up the correlation between electrospinning parameters and nanofiber morphology, patch architecture, molecular, and supermolecular structure as well as mechanical properties influencing the cellular response during in-vitro study. It is not possible to refer to all the literature on this subject; however, we show possible ways of tailoring the aforementioned properties by process parameters. This review is concentrated on pure polyesters and bicomponent nanofibers with Ge or Col. Blends: PCL/Ge, PCL/Col were selected as those which are considered very prospective in regenerative medicine and used in 3D scaffolds, wound dressing, drug delivery systems [21, 22]. At the end of the chapter, we try to answer the question which parameters are critical in the electrospinning process from the point of view of the morphology, structure, and properties of nanofiber patches influencing their functionality in tissue engineering.

2. Electrospinning process

Rayleigh was the first scientist who observed and described the electrospray process [23]. He discovered that a droplet in high voltage splints into small droplets through voltage gradient. This effect is named as "Rayleigh's instability". Nowadays, Zeleny [24] and Dole [25] continued investigations of Rayleigh. Dozin noticed that droplets from the electric field resemble aerosols in their external look (superficies) and properties [26]. Finally, electrospinning process was developed as a result of investigations on the electrospray process. The interest in electrospinning increased significantly after patent publishing in 1934 [27], which described the formation of cellulose fibers. In 1990s, PET nanofibers were electrospun, outlining new directions of scientific investigations and applications [28, 29].

Yarin and Reneker analyzed physics of the process and modeled the impact of parameters of the process on the jet move and formed fibers morphology [30, 31]. Those investigations enable understanding and controlling of the electrospinning process. Theoretical investigations of the electrospinning process were translated into use in nanotechnology, medicine, and texture industry by Kowalewski, Zussman, Wendorff, Ramakrishna and many others groups.

Every year, hundreds of publications and patents are published; according to Pubmed, for "electrospinning & tissue engineering", number of publication increased from 35 in 2006 to 262 in 2015; for "electrospinning & scaffolds", from 29 in 2006 to 235 in 2015 . According to available data [32], the highest number of issued patents is published in USA, China, South Korea, Japan, and Germany. It has hugely increased mainly in Europe (17%) and South Korea (7%). The top patent owners include companies such as DuPont (6%) and Donaldson (4%), and universities, with a major focus of the overall patent portfolio on the development of filtration media, methods, and apparatus. On the other hand, the process scale-up is still largely an issue even for electrospinning technologies, with a lot of space for further improvements. For example, from technological point of view, single length of fiber should be tailored/guided. In this case, electrospinning process exhibits poor reproducibility and accuracy in the production stage [33].

The equipment for the electrospinning process consists of high-voltage power supply, syringe with polymer solution, pump with controlled flow, and collector. **Figure 1** illustrates the schemas of vertical model of the setup.

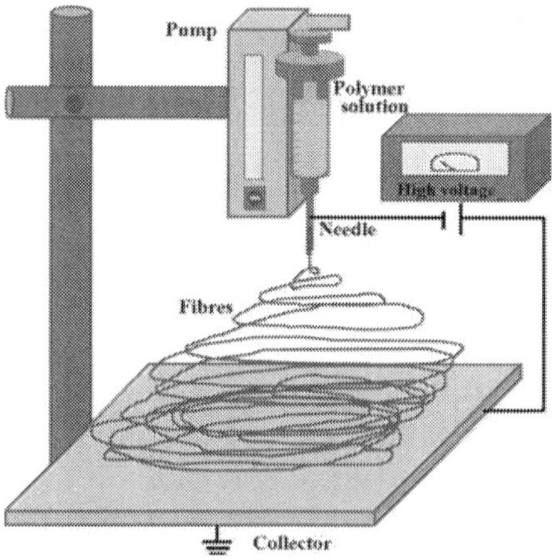

Figure 1. Schema of the process.

Depending on application, plates, drums, wires, or grids are used as the collector [10, 34]. To increase productivity of the process, multijet setups are used in industry and selected scientific labs [35]. Usually, home-made equipment is located in a laminar chamber, in rooms where temperature and humidity is controlled. Over the years, different methods of electrospinning have been considered: force spinning, multijet spinning, core-shell spinning, force-spinning, needleless-spinning, etc. (for example, see [10]).

The following companies offer commercial electrospinning supplies: Elmarco (www.elmarco.com), NaBond (www.electro-spinning.com), Holmarc Opto-Mechatronics (www.holmarc.com), E-Spin Nanotech (www.espinnanotech.com), Linari Engineering (www.linaribiomedical.com), Kato Tech (www.keskato.co.jp), Mecc Co. (www.mecc.co.jp), Toptec (www.toptec.co.kr), Electrospinz (www.electrospinz.co.nz), Electrospunra (www.electrospunra.com), IME Technologies (www.imetechnologies.nl), Yflow (www.yflow.com), and Inovenso (www.inovenso.com) [33].

3. Fundaments of electrospinning

Simplistically, high voltage is applied to the needle of the syringe with polymer solution. Electrostatic forces move the charges in the solution in the direction of the collector. Those forces exert effect on the solution droplet and deform it. Outflow of the solution (controlled by a dosing pump) and overrun of the critical value of the electrical field initialize flow of the solution into the collector. The solution is stretched under the high voltage into a thin jet. The solvent evaporates from the jet and formed fibers fall down on the collector [3].

Several steps can be distinguished during fiber formation in the electrospinning process (**Figure 2**):

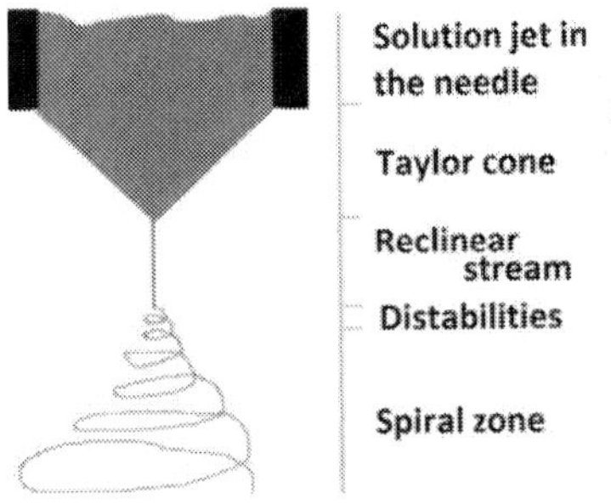

Figure 2. Zones during fibers forming in electrostatic field.

3.1. The solution in the needle

The first fiber forming zone, in the needle follows on initial orientation of the macromolecules solution to the transversal velocity gradient. In this case, the degree of orientation is determined by the ratio of the velocity gradient transverse to the rotational diffusion constant or the product of the gradient of the transverse flow rate of the solution and the relaxation time. Thus, the degree of orientation of the liquid jet leaving the needle increases with increasing flow rate and the length of the needle [36, 37].

3.2. Taylor cone

A drop of the solution after leaving the needle is exposed to an electrostatic field and takes the shape of a cone. Charge of the electric field is balanced by the forces of surface tension of the solution that changes it from a drop into a spherical cone. It creates a so-called Taylor cone at an angle of 49.3 degrees [38]. By the action of the electrostatic field, charges accumulated on the surface of the cone of the solution coexist with a higher charge density at the tip of the cone. The charge on the tip of the cone increases until exceeding the critical value at which the polarization reaches a value in excess of energy surface tension of the solution. The density of the charge at the top of the Taylor cone, which balances the surface tension of the polymer solution, is called the critical polarization *Vc*. After exceeding the critical value of polarization ($V > Vc$), the viscoelastic solution flows from the Taylor cone and there is a reduction in electrostatic charge density on the surface of the jet. Taylor showed that *Vc* can be described as [37]:

$$\left(Vc^2\right)^2 = 4\left(\frac{H^2}{L^2}\right)\left(ln\frac{2L}{R}\right) - 1.5)(0.117\pi R\gamma)$$

where *Vc* is critical polarization, *H* is distance of the needle to the collector, *L* is length of the needle, *R* is radius of the needle, and γ is surface tension of the solution.

After crossing *Vc*, the solution takes the form of a linear jet, being subjected to tension in the electric field (**Figure 3**).

Figure 3. Changes in droplet shape under different forces, where *V* is solution polarization and *Vc* is critical polarization.

3.3. Rectilinear jet of the solution

Zone of the rectilinear jet, leaving the "Taylor cone" is subjected to the electric field. Whereby the accumulation of electrostatic charges appears at the surface that causes the surface layer of the light polarization. In the layer of the surface of the solution, dipole (bipolar) layer electric is formed. This layer has always one type of charges facing toward the exterior surface and the other inside the jet.

Generated charged jet of the solution is defined by density of surface charges and electric potential. Aerial density of charges is varied along the length of the straight jet, due to Coulomb's law [39, 40].

In the fiber formation process, in a strong electrostatic field, the stretching of the jet reaches very high values. When the jet of the solution is straight, the stretching value is approximately 1000 times [39].

3.4. Disabilities in the rectilinear jet

Accumulated polarized charges on a straight stream interact with the generated electrostatic field and a specific wavelength stream is formed at the critical value of the electrostatic charge, initiating the formation of a large gradient flowing stream of the solution. Under these conditions, the release of internal cohesion trickles, and the solution begins to flow rapidly, causing turbulence of the straight section of the stream. The effect of turbulence is in the form of a cone. It changes the straight stream of traffic solution in a conical spiral motion that results from changes in the energy system. When the stream is subjected to bending, stretch reaches approximately 25,000 times [37].

3.5. The spiral movement of the solution and solvent evaporation

The spirals of the polymer solution are the result of exceeding the cohesive energy inside of the straight section of the stream by the value of the energy of the electrostatic field. As a result of these energy changes, the straight jet of the solution starts to form a cone of the spirals increasing in diameter.

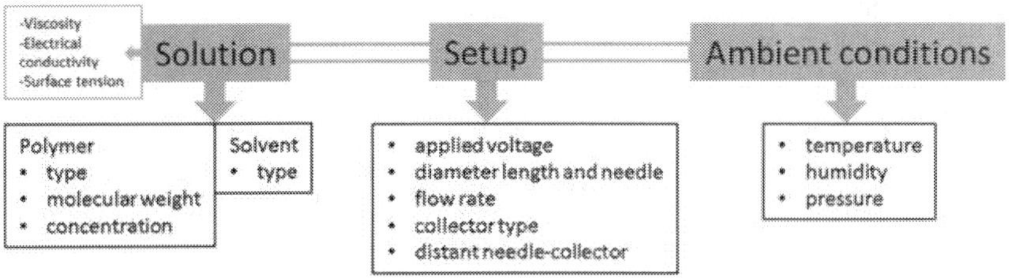

Figure 4. Influence of electrospinning process parameters on morphology and molecular structure.

Descriptions of the causes and effects of instability in a spiral cone spray can be found in many works by Reneker et al. [36, 41, 42].

About 15 main parameters of the process are distinguished. It is possible to divide them into three groups related with parameters of the process, setup, and ambient conditions (**Figure 4**).

Morphology describes diameter, shape, porosity, topography of the single fiber surface and also pillarization (architecture), fiber association, total porosity of the patches. Morphology of single fibers and patches is mostly described in literature by scanning electron microscopy. Fibers formed in the electrospinning process are characterized by a diameter of 50–2000 nm. The mean fiber diameter depends on the polymer, solvent type, and equipment parameters (**Figure 5**) [11, 31]. Dry fibers, which are uniform in morphology, are electrospun in a narrow range of process parameters, called a window process. All parameters of the process have to be optimized to form beads-free uniform fibers [44].

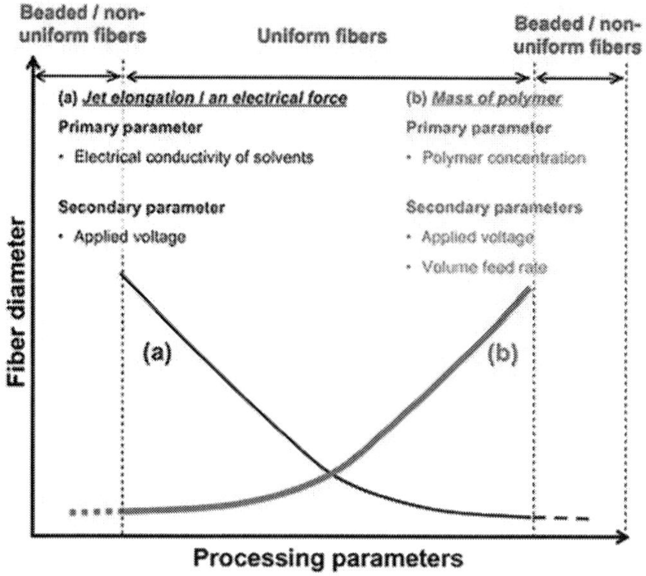

Figure 5. Fibers diameter as a function of the parameters related to (a) forces in solution jet and (b) polymer mass [10].

One of the first systematic descriptions of the influence of the process parameters on fiber morphology was prepared by Tan et al. [10].

In this simplification, parameters of the process were grouped into parameters which influence the jet elongation—electrical forces (blue line) and parameters which are related to the polymer mass (orange line). Each of them is divided into two individual groups: primary, exerting a stronger impact, and secondary, not playing such a significant role.

Primary parameters related to the polymer solution include electrical conductivity of the solvents and polymer concentration. Secondary parameters are related to applied voltage and flow rate. Generally, when impact of electrical conductivity of the solvent and applied voltage is lesser, thinner fibers are electrospun (smaller diameter); after achieving some maximum fibers with beads or only beads are formed. On the other hand, increase in polymer concentration, flow rate, and applied voltage contribute to smaller solution mass. As a consequence, thicker fibers are formed. Fibers with beads are formed when jet elongation is too effective or carried polymer mass is too low [45].

Deitzel et al. [45] reported that solution parameters had a stronger effect on fiber morphology than applied voltage. His observations overlapped with those of Ramakrishna et al., who found that too high applied voltage led to beads formation.

Results of the Thompson et al. model [31] indicate five parameters with a strong influence on the final diameter of the electrospinning jet, five parameters that have a moderate effect, and three parameters with only a minor effect on the fiber diameter. The most significant effect on the jet radius is exerted by volumetric charge density, distance from nozzle to collector, initial jet/orifice radius, relaxation time, and viscosity. The other parameters (initial polymer concentration, solution density, electric potential, perturbation frequency, and solvent vapor pressure) have a moderate effect on the jet radius. Parameters such as relative humidity, surface tension, and vapor diffusivity have a minor effect on the jet radius.

4. Materials

Biodegradable, synthetic polymers such as PCL, polylactic acid (PLA), polyglycolic acid (PGA) and their copolymers, less frequently used poly-3-hydroxybutyrate (PHB), polyurethane (PU) and natural polymers (biopolymers) are used in tissue engineering [4, 36]. Natural polymers like Col, Ge, Chit, elastin, silk, and keratin are commonly investigated. During the last few years, blends of synthetic polymers and biopolymers, or nanocomposites with nanoparticles (Au, Ag, hydroxyapatite [41, 46], carbon nanotubes [42], etc.) have been formed to combine their properties and enhance functionality. To enhance biological properties, the fiber surface is modified by plasma, chemical assembly, growth factors, and drugs. Another trend is addition of encapsulated cells, viruses [47].

This chapter concentrates on an analysis of literature about electrospun polyester and blends of polyesters with biopolymers: Ge, Col, Chit. Polyester properties and chemistry are profoundly described in literature [33, 44].

Col is the most commonly used natural material for cartilage regeneration. In the native tissue, it represents 90% of the cartilage skeletal tissue. Col is a natural polymer with a triple helical structure which contains sequences enabling cell adhesion, migration, and differentiation [48]. Additionally, Col type II has a high amount of bound carbohydrate groups, allowing Col scaffold-water interaction. Col can be easily formed into various scaffold architectures. Most common forms analyzed and applied are gel [37, 43], sponges [49, 50], and fibers [51]. There are several scaffolds dedicated to tissues regeneration made from bovine Col type I (Ne Celsius degreesart, Histogenics Corporation, Waltham, MA) or rat-tail Col type I (CaReS-1S, Arthro-Kinetics, Esslingen, Germany) [52].

Ge is a natural polymer, a denatured form of Col. It is obtained by controlled hydrolysis of Col. Two types of Ge can be distinguished: Type A—extracted and processed by acidic pretreatment from Col and Type B—obtained by alkaline pre-treatment [53]. Ge has been widely used because of its biological origin, biodegradability, biocompatibility, and commercial availability at a relatively low cost. The only disadvantage of Ge is its solubility in water without special treatment like cross-linking [54]. Nowadays, it has been used as sealant for vascular prostheses [55, 56], in drug delivery systems [33, 57, 58] or dressings for wound healing [59, 60]. Fiber and sponge forms are also obtained by various methods. Nevertheless, mixing Ge with other polymers, mostly synthetic polymers, is common practice [22, 61]. Ge is already commercially used in wound dressings and as hemostatic sponges as well as bone scaffold (Comfeel Plus, Duoderm, Granuflex, Ultec, 3M Tegaderm Hydrocolloid; Gelfoam is also used [62].

Chit is a semi-crystalline polysaccharide polymer. It is produced commercially by deacetylation of chitin from skeletons of crustaceans (crabs, shrimp) and cell walls of fungi. Chit is found at least in 60% of deacetylated chitin [63]. Due to the high structural similarity to glycosaminoglycans, which build ECM of the cartilage tissue, it is commonly used in scaffolds for cartilage regeneration [64]. Its degradation products: chondroitin sulfate, dermatan sulfate, hyaluronic acid, keratin sulfate, and glycosylated type-II Col are nontoxic and they are easily involved in synthesis of articular cartilage. Positive charge of Chit macromolecules is responsible for some problems with electrospinning uniform fibers from its solutions. On the other hand, the positive charge of D-glucosamine residues in Chit helps in interaction with negatively charged compounds from the cell membrane, which has effect on analgesic activity and inhibition of RNA synthesis of microorganisms [65].Chit is commercially used for cartilage regeneration and wound healing [66].

5. One component fibers

Fibers electrospun from one polymer type are analyzed as a prospective material for tissue engineering applications. This summary concentrates mostly on PCL, as one of the most biocompatible, biodegradable, and promising polymers used in regenerative medicine. However, many papers about poly(ethylene oxide) (PEO) are also mentioned because it was the main demonstrator of the relationship between parameters of the electrospinning process and properties of the fibers formed.

5.1. Solution parameters

Type, concentration of the polymer, and solvent type determine the solution properties (density, boiling point, dipole moment, etc.). Those properties like viscosity, electrical conductivity, surface tension are critical for morphology and structure of electrospun single fibers and fiber patches.

Solution viscosity is highly related to polymer molecular chains entanglement; low polymer concentration and/or low molecular weight yield lacks chains engagements. As a consequence, lack of polymer chains in the jet attacks in opposition to surface tension attacks in opposition to surface tension. Because of that, beads (so-called Rayleigh instability) are formed. Critical chains engagement is necessary to form uniform fibres.

Solution viscosity depends on the polymer type, its concentration, and solvent type. Taking into account manure parameters, it is possible to obtain appropriate solution viscosity for effective electrospinning process [3, 67]. Fong was one of the first people who described influence of viscosity on morphology of PEO fibers formed from aqua solution [12]. Beads were observed below defined viscosity, number of beads decreased with an increase in viscosity; the shape of beads changed from globular to spherical. After overrunning some critical value, bead-free fibers are formed. The diameter of uniform fibers changes with the polymer concentration.

The influence of the polymer concentration of aqueous PEO solutions on solution viscosity, surface tension, and diameter of electrospun fibers was evaluated [49]. Surface tension of the solution decreased with the polymer concentration, in opposition to the viscosity increase. However generally, the mean diameter of fibers increased with polymer concentration.

Molecular weight of a polymer also plays a significant role in formation of uniform fibers. PVA fibers electrospun from different molecular weight were analyzed. Electrospinning of low molecular weight PVA decide about electrospray forming and/or bead fibers formation [68].

Influence of molecular weight and polymer concentration on chains engagement was analyzed for some polymers and solution entanglement number (ne)s was determined [69].

$$(\mathrm{ne})s = \mathrm{Mw}/(\mathrm{Me})\mathrm{s}$$

where (Me)s is the entanglement molecular weight in solution, Mw is the weight-average molecular weight.

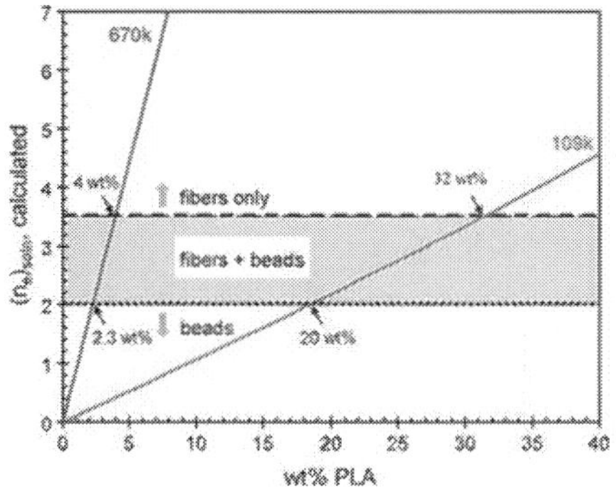

Figure 6. Fibers electrospinning in relations to solution entanglement number (ne)s and Mw in function of solution concentration (a) PLA and (b) PS [70].

Horizontal lines electrospray (*beads*), fibers with beads (*fibers + beads*) and fibers (**Figures 6a** and **6b**). Crosscut of the diagonal lines related to Mw of the polymer (blue, red, black, green) with horizontal lines (ne)s determines range of bead-free electrospun fibers. Examples: Mw 670kDa PLA below 4%, Mw 109kDa PLA below 32% uniform fibers were electrospun (**Figure 6a**).

PLA was electrospun from solutions differing in dichloromethane (DCM) concentration [71]. Viscosity and surface tension of the solution increased with polymer concentration; however, a decrease in electrical conductivity was recorded. Average fiber diameter and standard deviation of the diameter increased with the solution concentration. Analyses of electrical conductivity by pyridine additives were also performed.

Solution surface tension depends on the surface tension of the solvent, polymer concentration, and polymer molecular weight. Surface tension of the solution associated with low polymer concentration is usually higher than polymer chains engagement. Uniform distribution of the electrical charges possibly due to proper surface tension and viscosity (polymer concentration) enable formation of uniform fibers, free of beads [72]. Fiber diameter increases as a function of polyesters concentration. Electrical conductivity, dielectric constant, and viscosity enhance the effectiveness of forces applied to the solution [43].

Solutions characterized by relatively high electric conductivity carry more electrical charges during the electrospinning process. High charges density leads to effective stretching of the jet through repulsion of one sign charges. A highly conductive solvent eliminates high voltages during the process (deranging the process stability). Addition of DMF increases the solution conductivity; however, after exceeding a certain volume, beads are formed in the electrospinning process [73]. It is possible to enhance electrical conductivity by addition of polyelectric polymers with high electrical charge, like Ge and slats (e.g., NaCl).

Dielectric constant supports formation of uniform fibers; however, similarly to electrical conductivity, too high dielectric constant leads to beads formation.

Windows process for some polyesters has been described in literature [3, 19, 73].

Usually, round cross-section, ribbons or branched fibers are electrospun under certain conditions [65, 74]. The phenomenon of formation of ribbon fibers was described using gelatin as an example. During rapid evaporation of the solvent from the surface of the jet, the skin is formed; under air pressure fibers are pressed and ribbons are formed [42].

On the surface of electrospun fibers, pores are formed. They influence chemical and physical properties (tension, wettability). Surface porosity arises from the solvent type or ambient conditions [75].

Fewer descriptions of the influence of electrospinning parameters on crystallinity and molecular structure have been found than on morphology.

Crystallinity of electrospun fibers and casted films was determined from wide-angle X-ray scattering (WAXS) data by Lee [76]. A higher degree of crystallinity was estimated for casted films than for fibers. Intensity of peaks from PCL crystals (I from 110/I from 200) of electrospun Fibers was compered to peaks from casted films. Additionally, radial profiles of WAXS indicate partial orientation of PCL crystals along the fiber main axis in case of electrospun fibres.

Crystallinity was also diversified by different types of solvents. Thermal analyses, such as differential scanning calorimetry (DSC) and Raman spectroscopy, revealed lower crystallinity of PCL fibers electrospun from chloroform than from hexafluoroisopropanol (HFIP) [19]. The diameter of electrospun fibers, determined by properties of solvents, was not taken into account.

Influence of average Mw on molecular orientation was characterized using PVA as an example [77]. WAXS data indicate higher molecular orientation of PVA fibers electrospun from polymers characterized by higher molecular weight. It correlates with increase of crystallinity of fibers electrospun from PVA with higher Mw.

By optical microscopy (dark field), optical birefringence of PEO electrospun fibers was determined [78]. Degree of birefringence was not analyzed. In another publication, ordering of polymer macromolecules in the direction of the electric field was found in the needle [79].

Morphology and molecular structure of a single electrospun fiber influences mechanical properties.

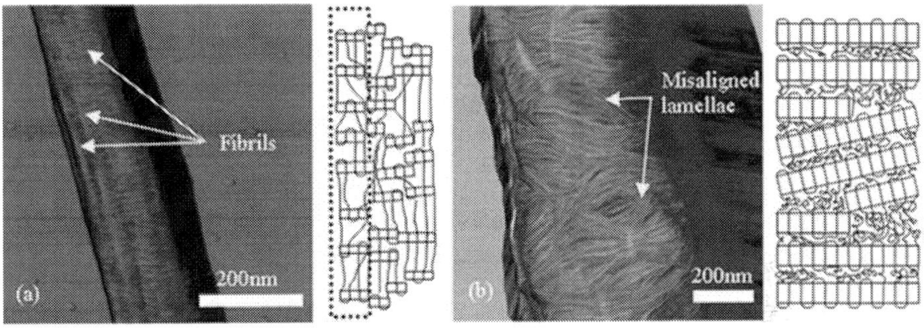

Figure 7. PCL fibers illustrated by AFM, electrospun from 10% solutions (a) 150 nm and (b) 450 nm [81].

Nanometric PCL fibers indicate higher Young's modulus than submicron ones [80]. It was observed that thin fibers (electrsopun from solution with low concentrations) were characterized by higher crystallinity than submicron fibers. It is opposite to publications which suggest increase in crystallinity with polymer concentration (fiber diameter). Authors suspect increase in molecular orientation with fiber diameter. Fibril structure of nanometric fibers and lamellar structure of submicron fibers was illustrated by atomic force microscopy (AFM) (**Figures 7a** & **Figures 7b**). Fibrils consist of staggered crystal and amorphous phases; lamellas consist of densely located lamellas and non-oriented amorphous phase.

Young's modulus of single fibers was analyzed in nanoindentanion mode by AFM for PLLA, PCL, Col fibers [82]. Analyses indicate higher Young's modulus of nanometric fibers than submicron ones [83]. It can be related to the higher order of chains in thin fibers and smooth surface.

Young's modulus decreases with fiber diameter, the increase in fiber diameter is observed for fibers electrospun from PCL solutions with concentration from 10–14 wt% [81]. In the same publication, significant elongation at break increase is observed with an increase in polymer concentration. These observations are confirmed on PVA [84]. Young's modulus determined from AFM in nanoindentation mode indicates decreased stiffness in fiber diameter. The analysis was performed for solution with 8% concentration. For polyvinylidene fluoride (PVDF) and PCL fibers, a decrease of Young's modulus as a function of diameter was observed [85].

5.2. Setup parameters

One of the most significant parameters in electrospinning process is applied voltage. Applied voltage regulates charges to the solution droplet. Density of the applied voltage, as value of applied charges on the surface of the droplet is described by applied voltage, distance between the needle and collector, and solution conductivity. It has to be related to the solvent type and polymer concentration.

Demir et al. [39] used polyurethane fibers and noticed that jet diameter increased with applied voltage; however, threshold voltage (necessary for start of the process) strongly depends on solution concentration [86].

Influence of applied voltage on the fiber diameter was found for PCL fibers electrospun from HFIP solution. Fiber diameter increases with same maximum value for fibers electrospun with different applied voltages in range of 7–20 kV [40].

Description of the applied voltage influence on crystallinity was not found for polyesters. Crystallinity as a function of applied voltage was found for cellulose fibers electrospun on plate [11]. Analyses indicate an increase of crystallinity with voltage, followed by a decrease of crystallinity after achieving a certain maximum value. It is explained as a result of powerful dynamics of crystallinity in high voltage which falters after the critical voltage is exceeded by longer distance fibers falling down.

Volume expansion of the solution is regulated by the flow rate settled by a medical pump. Flow rate is critical in Taylor cone formation; as a consequence of too fast flow rate, beads are formed, in the case of too slow flow rate, the needle clogs [15, 70].

Length nozzle optimization was also described. With the length of needles, polymer chains orientation increases in the jet and standard deviation of diameter of electrospun fibers decreases [87].

An appropriate distance between the needle and collector enables solvent evaporation which also strongly influences the fiber morphology (single, not agglomerated fibers).

Fibers from poly(hydroxybutyrate-co-hydroxyvalerate) (PHBV) were electrospun using positive applied (PV) and negative applied voltage (NV) [88]. In both PV and NV, it was found that the fiber diameter generally increased with increasing the solution concentration and increasing inner diameter but decreased with increasing the working distance, while the solution feeding rate did not significantly affect the fiber diameter. However, the fiber diameter increased with increasing PV but decreased with increasing NV. Polarity of the applied voltage did not significantly affect the water contact angle, which suggests no influence on the chemical structure of fibers. In the case of polyamide 11, changes of surface chemistry were observed while comparing fibers electrospun using NV and PV [76].

The architecture of electrospun patches depends on the collector type. Usually, fibers are electrospun on plates, drums, wires, and grids. Porosity of the patches may be changed by using different types of collectors [18]. Fibers electrospun on plate are of a random architecture, fibers electrospun on a drum are orientated in the direction of the drum rotation. Degree of fiber polarization is regulated by the speed of drum rotation and applied voltage [9, 89]. Fibers collected on drum have lower diameter than those collected on plate because of stretching during collecting (provided other parameters remain constant).

Porosity of electrospun fibers is within 60–92%, with the size of a single pore within the range of 1–465 μm [90].

Intensity of peaks registered by WAXS of PLLA fibers change depending on polymer concentration and degree of stretching; increase in molecular orientation and crystallinity was found as a function of degree of stretching (comparison: 63m/min and 630m/min) [15]. Peaks intensity increases with concentration, additionally with fibers annealing.

Molecular orientation of PAN electrspun fibers increases with the speed of drum rotation of 0–12.3 rot/min. Tailoring the rotation speed makes it possible to electrospin fibers of demanded molecular orientation [91]. High-speed rotation also guarantees paralyzation of electrospun fibers.

Mechanical analyses of electrospun fibers and patches are performed using AFM and traditional stress-strain equipment like INSTRON. The latter is more related to whole patches than single fibers. Typically, Young's modulus and maximum strain are determined. Differences in calculation arise from consulting real fibers amount on the sample cross-section.

In general, fibers collected on a drum display higher strength than those collected on a plate, which is related to molecular orientation of fibers collected on a drum [40].

5.3. Ambient conditions

Temperature plays a significant role during solvent evaporation. Time of crystallization decreases with the rising temperature which correlates with a weaker degree of crystallinity in the shorter time [3, 40]. Changes of temperature also lead to fiber porosity [15].

Air humidity has an effect on electrical conductivity of the ambient environment. Fiber diameter decreases with humidity when we change it from 30–50% [3]. Relatively low applied voltage and high viscosity can be used in 50% of humidity. Changes of humidity also influence surface porosity of a single fiber [92].

Usually, the electrospinning process takes place in air atmosphere; however, electrospinning in atmospheres of different gases is possible.

Temperature and humidity are not strictly controlled, as a consequence not so many researches describe influence of these parameters on fiber morphology and structure.

6. Bicomponent fibers with Col, Ge, and Chit

In the last few years, some literature has confirmed positive influence of biopolymer additive on the functionality of electrospun patches in tissue engineering applications. Most common biopolymers added to synthetic polymers are Col, Ge, and Chit.

6.1. Fibers with Col and Ge

Biopolymer additives to polyester influence the morphology as well as molecular and supermolecular structures of electrospun fibers. This part of description concentrates on blends with Ge and Col. Synthetic/biopolymer w/w% ratio is a subsequent parameter of the solution parameters group.

A decrease in fiber diameter was observed as a function of applied biopolymer or poly-lactic-co-glycolic acid (PLGA)/Ge and poly(lactide-co-ε-glycolic acid) (PLCL)/Ge blends [93]. Additionally, smaller fiber diameter and pore size of the patches electrospun on a drum in comparison to fibers collected on a plate were seen.

From the viewpoint of tissue engineering application (in-vitro and in-vivo study), miscibility (compatibility) of both polymers or interactions of their chemical groups seem to be important. Literature about miscibility of biopolymer and polyesters is sketchy and inconsistent.

Phase separation of Ge/PCL (1:1) in tetrafluoroethylene (TFE) and its consequences on electrospun fibers were analyzed by Feng et al [81, 94] and continued by Xue et al. [95]. Fluorescence observation of the FITC-conjugated Ge/PCL/TFE solution and dynamic light scattering (DSL) methods confirm the phase separation in the mixed solution. Phase separation leads to electropsun fibers splashing, bonding, and distribution in fiber diameter. Small addition of acetic acid to solution was found to lead to the opaque solution becoming transparent immediately without occurrence of precipitation for over 1 week. Nanofibers formed from solution with acetic acid are thinner, smoother, and more uniform. Increase in tensile strength and decrease in strain at break in fibers electrospun without acetic acid was observed [96].

Miscibility of polymer depends on glass temperature Tg. It is known that Tg value of miscible blends is between the values of those for both components [97]. In immiscible blends, two separate Tg values are observed [98].

Tg is usually determined from DSC data. However, it depends on the ambient temperature and humidity, so changes are not always connected exclusively with both components' miscibility. Euna et al. [99] describes strong influence of water on Tg of Ge. Tg of some blends within effect from crystal melting of one component PCL/Ge (Tg of Ge about 48°C), which impedes miscibility analyses from Tg.

Zhang et al. observed phase separation in PCL/Ge blends electrospun from TFE. As a consequence of leaching, pores on the surface of electrospun fibers appeared [100].

Phase separation was observed for PLLA/Ge in the analyses. Observation by transmission electron microscopy (TEM) illustrated Ge phase located along fiber axes (**Figure 8**) [86].

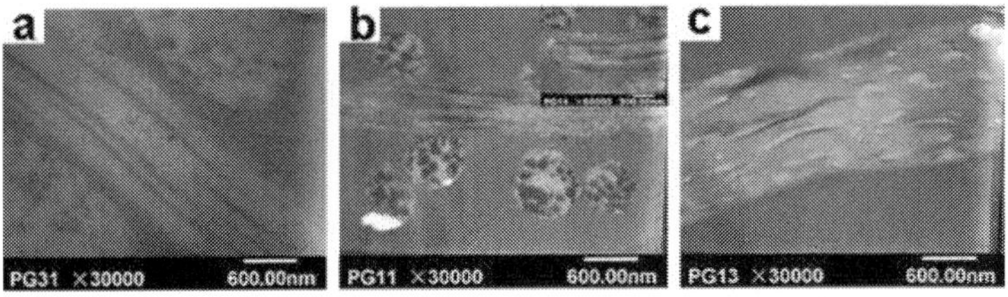

Figure 8. TEM observations of PLLA/Ge. (a) 3:1, (b) 1:1, and (c) 1:3 [101].

Fourier Transform Infrared Spectroscopy (FTIR) analyses of the Chit-PEO fibers indicate modified spectrum (between both polymers) [102].

In the article, *Electro-spinning of pure collagen nano-fibers – Just an expensive way to make Ge?*, polar solvent denaturants helixes structure of peptides are shown. HFIP, TFE probably destroy weak hydrogen bonds related to helixes structure [103].

DSC curves for PCL in PCL/Ge blends indicate overlapping of signals from both components (PCL,Ge) and dependence of PCL crystallinity on solvent type [19]. Crystallinity of PCL in PCL/Ge 50:50 from HFIP and TFE data was found to be 0.20 and 0.25, respectively. Miscibility of PCL/ Ge blends analysed from WAXS data was proved by shifts of peaks from PCL crystals (shifts of 110, 200).

Molecular orientation also changes with addition of Ge to PCL, as was observed on drum-collected fibers [79]. PCL crystallinity increases in the presence of a small amount of Ge.

In patches functioning as 3D cells scaffolds, wettability plays a significant role. Polyesters are hydrophobic, the contact angle of PCL is 109° [104], Ge 76° [105], Col 50°. PCL/Ge, PLGA/Ge blends show higher wettability than pure polymers, the contact angle is within 0–20° [72]. Lack of additives of this parameter has not been clearly explained in literature. In some publications, it is described as a result of ordering nonpolar groups of polymer in proximity of synthetic polymer; in some others, ionization of Ge is considered to play a significant role.

6.2. Fibers with Chit

Solution prepared from pure Chit and 90% acetic acid showed an increase in viscosity with higher Chit concentration. Uniform fibers are easily electrospun for Mn 106 kDa; for 30 kDa, the engagement of chains was too small, so beads were formed, for 398 kDa, driven mass was too high to be driven by applied voltage, so again beads were formed [106]. Electrospinning window process of Chit dissolved in acid aqueous solution was determined by Vrieze et al. [107]. Optimal values of electrospinning parameters to form uniform fibers are Chit concentration of 3%, concentration of acetic acid of 90%, collector-needle distance of 10 cm, solution flow rate of 0.3cml/h, applied voltage of 20 kV. Use of these parameters leads to the formation of uniform fibers, with a diameter of 70 ± 45 nm.

In the case of PCL/Chit blends, an increase of conductivity was observed with an addition of Chit to PCL [108], as a consequence, an increase in the diameter of fibers electrospun from blends in comparison to those from pure PCL was observed.

Chen et al. analyzed changes of diameter as a function of PLCL/Chit (1:1 mol %) concentration and ratio of PLCL/Chit [109]. Analyses indicate a decrease in the fiber diameter in the first case, and a decrease in diameter with aggregate polymers concentration. Increase of porosity and pore size was observed with Chit to PLCL additive. The porosity of patches electrospun from blends decreased in the case of collection-oriented fibers (drum collector) in comparison to fibers collected randomly on a plate or grid [131].

Peaks observed on FTIR spectra of PCL/Chit fibers, at 1675 and 1530 cm^{-1} correspond to the stretching of amide and protonated amine groups resulting from residual trifluoroacetic acid (TFA), as was also reported by Sangsanoh and Supaphol [110]. Spectra indicate that there is no miscibility between PCL/Chit chains, so no covalent bonding between the polymers occurred [111, 112]. The authors suggested that only physical interactions, such as the intermolecular hydrogen bonding between the carbonyl group of PCL and hydroxyl or ammonium ions of Chit between the Chit and PCL, remained. PCL/Chit fibers incubated in cell culture media for two weeks did not swell. This capability of retaining structural integrity in an aqueous environment confirms interaction of chains, possibly by engagement of Chit chains by PCL macromolecules.

XPS data confirm differences in surface analyses of blend electrospun using NV and PV applied to the needle [107].

Fibers electrospun from pure Chit are fully amorphous; polyesters and copolymers usually display crystallinity. WAXS observations indicate that blends elctrospun from PLCL/Chit (1:1) exhibit less crystallinity than pure electrospun PLCL. It suggests that Chit destroyed most of the regular arrangements of PLCL [131].

In the case of PCL/Chit blends, the intensity of the higher peak tends to decrease with the increase in the Chit weight content [113]. It is also evident that the peak position shifts to a smaller angle with the addition of Chit. It has been demonstrated that the conjugation of PCL with Chit suppresses the crystallization of both Chit and PCL to a certain extent. It has also been suggested that Chit and PCL chains are well mixed at a molecular level.

Decrease in tensile stress and ultimate strain and stiffness with Chit to PLCL [131], PCL addition was observed [135].

Pure Chit is hydrophilic (contact angle of about 67°, while polyesters are usually hydrophobic. Contact angle of PCL/Chit blend fibers increases from 109° for PCL/Chit 3:1 to 82° for 1:3 blend [135].

7. Biological investigation on one- and bicomponent electrospun fibers

In general, the product dedicated to use in tissue engineering should provide a repeatable and long lasting effect on treated tissue. Because of that, scaffolds should fulfill the following conditions connected with their functionality and the manufacturing process:

a. Biocompatible and bioresorbable;

b. Three-dimensional (3D) form which encourages formation of tissue;

c. Highly porous structure; interconnected/permeable pore network;

d. Mechanical properties suitable for application;

e. Controllable degradation rates;

f. Surface chemistry suitable for cellular attachment, proliferation; good connection with the surrounding tissues;

g. Production of scaffold should be easy and give repeatable results;

h. Prepared product should be easy in application; and

i. Price of the product should be affordable.

Electrospinning process is flexible enough to fulfill all of these requirements. Fibers from biocompatible, biodegradable polymers are easily prepared; additionally, it is possible to tailor their morphology, architecture, mechanical properties, etc., by electrospinning process parameters (paragraph 2–3). Electrospun patches exhibit many similarities in architecture and porosity degree to natural ECM.

Influence on fiber morphology modulated by PCL concentration and applied voltage was described [11]. PCL fibers electrospun from 8, 15, 20% HFIP solution exhibit increased fibroblast proliferation which correlates with an increase in diameter as a function of concentration. Influence of applied voltage changes in time: increase of proliferation with voltage after 3 days, increase after 6 days. Negative influence of beads on cell proliferation was observed.

Analyses of cellular response indicate a complex relationship between the diameter of fibers electrospun from PCL tailored by solution concentration (8–20%) and cell proliferation [96]. Beaded structures prevent cellular proliferation, indicating the need for uniform fibers for tissue-engineering scaffolds. Maximum positive effect of fiber diameter was observed for fibers with a diameter of 400 μm.

Electrospun patches showed a relatively small size of single pores. Pores facilitate cell migration and transport [114]. Porosity of PCL patches can be expanded by appropriate collectors up to 10 times [90]. Cell migration increases with total porosity and pore size, unfortunately publications about the degree of changes were not found.

Some tissues display a parallel arrangement of cells and ECM fibers (muscle, heart). It was the impulse for the formation of scaffolds with parallel arrangement of fibers. Drums and parallel wires are used for that purpose. For instance, in tendons and ligaments, stretch forces are 200–500 times higher in the fiber direction (forces generation) than perpendicular to this direction [115]. In cartilage joints, forces on the core are higher on the surface, and the highest in the direction of the main Col fiber [99]. It is another reason to analyze and mimic the nature; core fibers are formed as a consequence.

Increase of proliferation and migration was observed in PCL patches with a single pore size enhanced to about 50 μm [116]. Additionally, cell migration was supported by *"layer by layer"* technique, in which fiber spinning and cell seeding is held layer by layer.

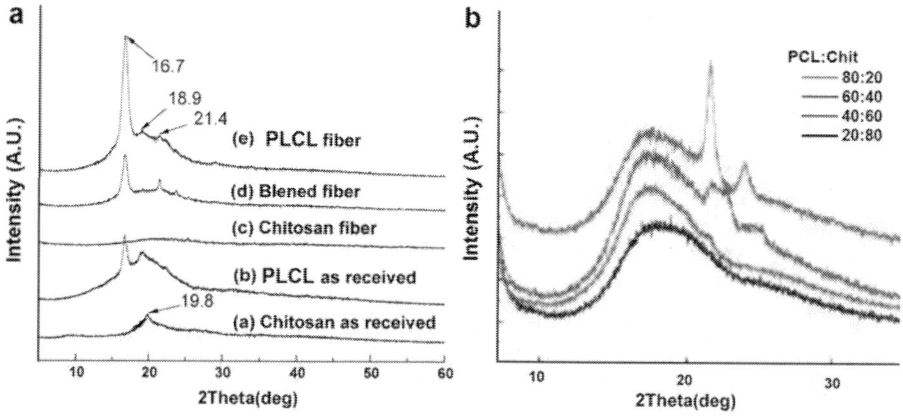

Figure 9. WAXS radial profile of (a) PLCL, Chit and PLCL/Chit 1:1 [131] and (b) PCL/Chit blends with different ratios [135].

Schwann cells seeded on PCL patches indicate parallelization in the direction of fibers. It is the sign of interaction scaffolds–cell, described in literature as a contact guidance (**Figure 9**) [101].

Analyses of the polyester/biopolymer blends influence on patches functionality are usually related to the arbitrally adopted ratio of the biopolymer about 20/30%.

The only publication found describes PLCL/Ge patches, electrospun from solutions of 100:0, 90:10, 70:30, 50:50, 30:70, and 0:100 [113]. A decrease in fiber diameter was observed as a function of the Ge amount (**Figure 10**), increase in total porosity (from 52 to 74%) and Young's modulus. Miscibility of PLC/Ge was determined from DSC data, and it was found that 90:10 was miscible, while other blends were immiscible.

Neural stem cells (C17.2) seeded on PCL, PCL/Ge 70:30 patches indicate higher activity determined by mitochondrial activity test after 2, 4, and 6 days on drum-collected patches (oriented fibres) than on plate (random fibres) [117]. It is described as a result of high hydrophilicity of PCL/Ge patches and contact guidance on oriented patches–cells skeleton elongated along the fiber direction.

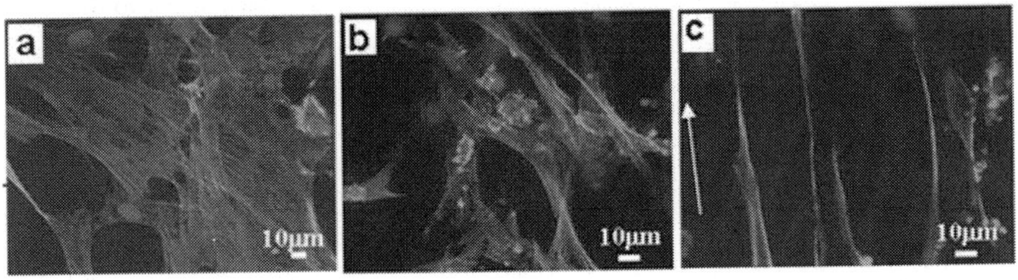

Figure 10. Observations by fluorescence microscopy after 7 days, cells seeded on (a) control, (b) non-oriented fibers, and (c) oriented (parallel) fibers [142].

Nerve cells seeded on PLCL/Col patches indicate higher activity on patches with Col type I and III than with Col I additive. Contact guidance on oriented patches was observed [5].

Cancer cells seeded on PCL, PCL/Ge patches were also analyzed by Hartman et al. [19]. Addition of Ge to PCL supports functionality, as scaffolds are more effective when HFIP is used as a solvent than TFE. A decrease in diameter and increase in Young's modulus were observed as a function of additive.

Decryption of cells differentiation on polyester blends with Ge or Col is picked up rarely. However, stem cells differentiation to nerve is more effective on PCL/Col patches than with those polymers electrospun separately [118]; (PCL/Ge 50:50 patches were taken into account).

PCL/Ge 7:3 blend electrospun on a drum displays higher Young's modulus and less mass lost during 12 weeks degradation than patches electrospun on a plate (random) [119]. Proliferation degree was higher on fibers electrospun on a drum, which correlates with higher Young's modulus.

Pure Chit electrospun from 70:30 v/v TFA/DCM (trifluoroacetic acid/dichloromethane) was analyzed as a substrate toward Schwann cells, osteoblast, keratinocyte, and fibroblast cells viability, proliferation spreading [117]. Viability and proliferation of osteoblasts cultivated on Chit fibers was highest in comparison to TCP and other cells types. Osteoblasts were also the most spread in comparison to the other cells analyzed.

Human adipose-derived stem cells (hASCs) indicate better proliferation on PCL/Ge 1:1 with addition of bone powder than on pure PCL/Ge [120].

That indicates that additional natural bioactivator enhances the cellular response. PCL/Chit blends, containing various amounts of Chit (0, 3, 9, and 23%) were electrospun and evaluated for their osteogenic differentiation of preosteoblasts [135]. Osteoblasts cultivated on PCL/Chit exhibit calcium deposition, alkaline phosphatase (ALP) activity, and the expression of osteopontin (OPN). Culture of cell-rich 3D constructs prepared by assembling cells layer-by-layer on PCL/Chit nanofibers led to a uniform tissue formation with significant mineralization at 21 days.

Patches electrospun from PCL/Chit solution were designed and evaluated in-vitro using rat Schwann cells for nerve tissue engineering [121]. Investigations indicate smaller mean diameter of blended fibers than pure polymers as follows: 630, 450, and 190 nm, respectively, for PCL, Chit, PCL/Chit blend, where PCL/Chit scaffolds fabricated using an electrospinning process showed better cell proliferation than PCL scaffolds and maintained their characteristic cell morphology, with spreading bipolar elongations to the nanofibrous substrates.

Post-treatment by cross-linking of PCL–gelatin, PLGA-gelatin, PLGA–Chit tubes was analyzed in case of cytotoxicity [122]. All patches showed about 5% increase in toxicity after cross-linking by glutaraldehyde. However, other analyses confirm higher liquid strength of cross-linked scaffolds than non-cross-linked scaffolds. It was a consequence of superior mechanical properties of a cross-linked scaffold and a decrease of pore size.

8. Conclusions

Nowadays, fibers formed in electrospinning process are commercially available in many branches of industry, including regenerative medicine and tissue engineering. Electrospinning is a success story, one of the very few belonging to the broad area of nanotechnologies which hold so great, unprecedented promises. Understanding of electrospinning process parameters enables tailoring of electrospun nanofibers morphology, internal structure, and properties. This opens up new possibilities for the use of fibers and improves the solutions already produced on the basis of electrospun fibers.

Mammal cells are sensitive enough to respond even to small changes in the above parameters as well as to changes in the chemical structure. Cellular attachment, proliferation, differentiation occur at the surface of the fiber through the interaction of polymer molecules with the adhesion proteins (focal adhesion) found on the cellular surface. Use of electrospun nanofibers in tissue engineering is also driven by process flexibility and similarities in architecture to natural ECM. Different morphology, architecture, porosity, and alternative polymers with bio-activators are combined easily in the electrospinning process. Additionally, low cost of fiber production accompanies the aforementioned advantages.

Regulations applicable to scaffolds (medical devices) are restrictive, so the commercialization process is usually long in this case. However, tailoring of fiber composition appropriate for the application and cooperation with companies specializing in the industrial upscaling of electrospinning process are a guaranty of the success like in the case of Tegaderm, 3M Medical, TranCyte, Dermagraft-TC, Chito-Flex, Integra, etc., for wound dressing.

Acknowledgements

This work was supported by grants: 2013/09/D/ST8/04009 from the Polish National Centre of Sciences and 388/L-6/2014 from the The National Centre for Research and Development.

Author details

Dorota Kołbuk*

Address all correspondence to: dkolbuk@ippt.pan.pl

Institute of Fundamental Technological Research, Polish Academy of Sciences, Warsaw, Poland

References

[1] Luo CJ, Stoyanov SD, Stride E, Pelan E, Edirisinghe M. Electrospinning versus fibre production methods: from specifics to technological convergence. Chem Soc Rev. 2012;41(13):4708. DOI: 10.1039/c2cs35083a

[2] Cort JR, Andersen NH. Formation of a molten-globule-like state of myoglobin in aqueous hexafluoroisopropanol. Biochem Biophys Res Commun. 1997;233(3):687–91. DOI: 10.1006/bbrc.1997.6524

[3] Ramakrishna S, Fujihara K, Teo W-E, Lim T-C, Ma Z. An Introduction to Electrospinning and Nanofibers. World Scientific Pub Co Pte Lt; 2005. p. 396. DOI: 10.1142/9789812567611

[4] Kijeńska E, Prabhakaran MP, Swieszkowski W, Kurzydlowski KJ, Ramakrishna S. Electrospun bio-composite P(LLA-CL)/collagen I/collagen III scaffolds for nerve tissue engineering. J Biomed Mater Res. 2012;21;100B(4):1093–102. DOI: 10.1002/jbm.b.32676

[5] Doshi J, Reneker DH. Electrospinning process and applications of electrospun fibers. In: Proceedings of the IEEE Industry Applications Conference Twenty-Eighth IAS Annual Meeting; 2-8 October 1993; Toronto, Ontario. New York: 1993. p. 1698–1703. DOI: 10.1109/ias.1993.299067

[6] Langer R, Vacanti J. Tissue engineering. Science. 1993;14;260(5110):920–6. DOI: 10.1126/science.8493529

[7] Sell SA, Wolfe PS, Garg K, McCool JM, Rodriguez IA, Bowlin GL. The use of natural polymers in tissue engineering: a focus on electrospun extracellular matrix analogues. Polym. 2010;9:522–53. DOI: 10.3390/polym2040522

[8] Cui W, Zhou Y, Chang J. Electrospun nanofibrous materials for tissue engineering and drug delivery. Sci Technol Adv Mater. 2010;11(1):014108. http://dx.doi.org/10.1088/1468-6996/11/1/014108

[9] Bhardwaj N, Kundu SC. Electrospinning: a fascinating fiber fabrication technique. Biotech Adv. 2010;28(3):325–47. DOI: 10.1016/j.biotechadv.2010.01.004

[10] Tan S-H, Inai R, Kotaki M, Ramakrishna S. Systematic parameter study for ultra-fine fiber fabrication via electrospinning process. Polym. 2005;46(16):6128–34. DOI: 10.1016/j.polymer.2005.05.068

[11] Zhao S, Wu X, Wang L, Huang Y. Electrospinning of ethyl-cyanoethyl cellulose/tetrahydrofuran solutions. J Appl Polym Sci. 2003;91(1):242–6. DOI: 10.1002/app.13196.

[12] Salem DR. Electrospinning of nanofibers and the charge injection method, In: Brown PJ, Stevens K, editors. Nanofibers and Nanotechnology in Textiles. Cambridge: Woodhead; 2007. p. 3–21. DOI: 10.1533/9781845693732

[13] Hartman O, Zhang C, Adams EL, Farach-Carson MC, Petrelli NJ, Chase BD, et al. Biofunctionalization of electrospun PCL-based scaffolds with perlecan domain IV peptide to create a 3-D pharmacokinetic cancer model. Biomat. 2010;31(21):5700–18. DOI: 10.1016/j.biomaterials.2010.03.017.

[14] Ding B, Kim H-Y, Lee S-C, Shao C-L, Lee D-R, Park S-J, et al. Preparation and characterization of a nanoscale poly(vinyl alcohol) fiber aggregate produced by an electrospinning method. J Polym Sci, B: Polym Physics. 2002;20;40(13):1261–8. DOI: 10.1002/polb.10191

[15] Kim CH, g YH, Kim HY, Lee DR, Dharmaraj N, Choi KE. Effect of collector temperature on the porous structure of electrospun fibers. Macromol Res. 2006;14(1):59–65. DOI: 10.1007/bf03219069

[16] Cui H, Sinko PJ. The role of crystallinity on differential attachment/proliferation of osteoblasts and fibroblasts on poly (caprolactone-co-glycolide) polymeric surfaces. Front Mat Sci. 2011;7;6(1):47–59. DOI: 10.1007/s11706-012-0154-8

[17] Balasundaram G, Sato M, Webster TJ. Using hydroxyapatite nanoparticles and decreased crystallinity to promote osteoblast adhesion similar to functionalizing with RGD. Biomat. 2006;27(14):2798–805. DOI: 10.1016/j.biomaterials.2005.12.008

[18] Yaylaoglu M. Development of a calcium phosphate–gelatin composite as a bone substitute and its use in drug release. Biomat. 1999;20(8):711–9. DOI: 10.1016/s0142-9612(98)00199-9

[19] Woodruff MA, Hutmacher DW. The return of a forgotten polymer—Polycaprolactone in the 21st century. Prog Polym Sci. 2010;35(10):1217–56. DOI: 10.1016/j.progpolymsci.2010.04.002

[20] Sambudi NS, Sathyamurthy M, Lee GM, Park SB. Electrospun chitosan/poly(vinyl alcohol) reinforced with CaCO3 nanoparticles with enhanced mechanical properties

and biocompatibility for cartilage tissue engineering. Comp Sci Tech. 2015;106:76–84. DOI: 10.1016/j.compscitech.2014.11.003

[21] Kim M, Kim GH. Electrohydrodynamic direct printing of PCL/collagen fibrous scaffolds with a core/shell structure for tissue engineering applications. Chem Eng J. 2015;279:317–26. DOI: 10.1016/j.cej.2015.05.047

[22] Kołbuk D, Sajkiewicz P, Maniura-Weber K, Fortunato G. Structure and morphology of electrospun polycaprolactone/gelatine nanofibres. Eur Polym J. 2013;49(8):2052–61. DOI: 10.1016/j.eurpolymj.2013.04.036

[23] Rayleigh, Lord. XX. On the equilibrium of liquid conducting masses charged with electricity. Philosl Mag, 5. 1882;14(87):184–6. DOI: 10.1080/14786448208628425

[24] Zeleny J. Instability of Electrified Liquid Surfaces. Phys Rev. 1917;10(1):1–6. DOI: 10.1103/physrev.10.1

[25] Dole M. Molecular Beams of Macroions. J Chem Phys. 1968;49(5):2240. DOI: 10.1063/1.1670391

[26] Drozin VG. The electrical dispersion of liquids as aerosols. J Coll Sci. 1955;10(2):158–64. DOI: 10.1016/0095-8522(55)90022-2

[27] Anton F. *Process and Apparatus for Preparing Artificial Threads*. U.S. Patent No 1,975,504, 1934.

[28] Larrondo L, St. John Manley R. Electrostatic fiber spinning from polymer melts. II. Examination of the flow field in an electrically driven jet. J Polym Sci Polym Phys Ed. 1981;19(6):921–32. DOI: 10.1002/pol.1981.180190602

[29] Larrondo L, St. John Manley R. Electrostatic fiber spinning from polymer melts. III. Electrostatic deformation of a pendant drop of polymer melt. J Polym Sci Polym Phys Ed. 1981;19(6):933–40. DOI: 10.1002/pol.1981.180190603

[30] Yarin AL, Koombhongse S, Reneker DH. Bending instability in electrospinning of nanofibers. J Appl Phys. 2001;89(5):3018. DOI: 10.1063/1.1333035

[31] Thompson CJ, Chase GG, Yarin AL, Reneker DH. Effects of parameters on nanofiber diameter determined from electrospinning model. Polym. 2007;48(23):6913–22. DOI: 10.1016/j.polymer.2007.09.017

[32] Persano L, Camposeo A, Tekmen C, Pisignano D. Industrial upscaling of electrospinning and applications of polymer anofibers: a review. Macromol Mater Eng. 2013;8;298(5):504–20. DOI: 10.1002/mame.201200290.

[33] Iwata T, Yoshiharu D. Crystal structure and biodegradation of aliphatic polyester crystals. Macromol Chem Phyis. 1999;200;11:2429-2442. DOI: 10.1002/(SICI)1521-3935(19991101)200:11<2429::AID-MACP2429>3.0.CO;2-#

[34] Teo WE, Ramakrishna S. A review on electrospinning design and nanofibre assemblies. Nanotech. 2006;30;17(14):R89–R106. DOI: 10.1088/0957-4484/17/14/r01

[35] Srivastava Y, quez M, Thorsen T. Multijet electrospinning of conducting nanofibers from microfluidic manifolds. J Appl Polym Sci. 2007;14;106(5):3171–8. DOI: 10.1002/app.26810

[36] Liu CZ, Czernuszka JT. Development of biodegradable scaffolds for tissue engineering: a perspective on emerging technology. Mat Sci Tech. 2007;23(4):379–91. DOI: 10.1179/174328407x177027

[37] Kawamura S, Wakitani S, Kimura T, Maeda A, Caplan AI, Shino K, et al. Articular cartilage repair: rabbit experiments with a collagen gel-biomatrix and chondrocytes cultured in it. Acta Orthop Scand. 1998;69(1):56–62. DOI: 10.3109/17453679809002358

[38] Garg K, Bowlin GL. Electrospinning jets and nanofibrous structures. Biomicrofluidics. 2011;5(1):013403. DOI: 10.1063/1.3567097

[39] Demir M., Yilgor I, Yilgor E, Erman B. Electrospinning of polyurethane fibers. Polym. 2002;43(11):3303–9. DOI: 10.1016/s0032-3861(02)00136-2

[40] Kolbuk D, Guimond-Lischer S, Sajkiewicz P, Maniura-Weber K, Fortunato G. The effect of selected electrospinning parameters on molecular structure of polycaprolactone nanofibers. Int J Polym Mat Polym Biomat. 2014;2;64(7):365–77. DOI: 10.1080/00914037.2014.945209

[41] Ryszkowska JL, Auguścik M, Sheikh A, Boccaccini AR. Biodegradable polyurethane composite scaffolds containing Bioglass® for bone tissue engineering. Comp Sci Tech. 2010;70(13):1894–908. DOI: 10.1016/j.compscitech.2010.05.011

[42] Zhu Y, Li C, Cebe P. Poly(lactides) co-electrospun with carbon nanotubes: thermal and cell culture properties. Eur Polym J. 2016;75:565–76. DOI: 10.1016/j.eurpolymj.2016.01.014.

[43] Kimura T, Yasui N, Ohsawa S, Ono K. Chondrocytes embedded in collagen gels maintain cartilage phenotype during long-term cultures. Clin Orthop Relat Res. 1984;186:231-9. DOI: 10.1097/00003086-198406000-00039.

[44] Idaszek J, Bruinink A, Święszkowski W. Delayed degradation of poly(lactide-co-glycolide) accelerates hydrolysis of poly(ε-caprolactone) in ternary composite scaffolds. Polym Degrad Stability. 2016;124:119–27. DOI: 10.1016/j.polymdegradstab.2015.12.020

[45] Deitzel J., Kleinmeyer J, Harris D, Beck Tan N. The effect of processing variables on the morphology of electrospun nanofibers and textiles. Polym. 2001;42(1):261–72. DOI: 10.1016/s0032-3861(00)00250-0

[46] Auguścik M, Prociak A, Rojek P, Ryszkowska J. Polyurethane-urea substrates from rapeseed oil-based polyol for bone tissue cultures intended for application in tissue

engineering. Polym Deg Stab. 2014;108:241–9. DOI: 10.1016/j.polymdegradstab.2014.03.010

[47] Salalha W, Kuhn J, Dror Y, Zussman E. Encapsulation of bacteria and viruses in electrospun nanofibres. Nanotech. 2006;30;17(18):4675–81. DOI: 10.1088/0957-4484/17/18/025

[48] Kleinman HK. Role of collagenous matrices in the adhesion and growth of cells. J Cell Biol. 1981;1;88(3):473–85. DOI: 10.1083/jcb.88.3.473

[49] Speer DP, Chvapil M, Vorz RG, Holmes Md. Enhancement of Healing in Osteochondral Defects by Collagen Sponge Implants. Clin Orthop Relat; 1979;144:326-335. DOI: 10.1097/00003086-197910000-00056.

[50] Lu H, Ko Y-G, Kawazoe N, Chen G. Cartilage tissue engineering using funnel-like collagen sponges prepared with embossing ice particulate templates. Biomat. 2010;31(22):5825–35. DOI: 10.1016/j.biomaterials.2010.04.019

[51] Teuschl AH, Nürnberger S, Redl H, Nau T. Articular cartilage tissue regeneration—current research strategies and outlook for the future. Eur Surg. 2013;45(3):142–53. DOI: 10.1007/s10353-013-0217-9

[52] Irion VH, Flanigan DC. New and Emerging Techniques in Cartilage Repair: Other Scaffold-Based Cartilage Treatment Options. Operat Sports Med. 2013;21(2):125–37. DOI: 10.1053/j.otsm.2013.03.001

[53] Zhou P, Regenstein JM. Effects of Alkaline and Acid Pretreatments on Alaska Pollock Skin Gelatin Extraction. J 'Food Sci. 2005;70(6):c392–c396. DOI: 10.1111/j.1365-2621.2005.tb11435.x

[54] Zhang Y, Ouyang H, Lim CT, Ramakrishna S, Huang Z-M. Electrospinning of gelatin fibers and gelatin/PCL composite fibrous scaffolds. J Biomed Mater Res. 2004;72B(1):156–65. DOI: 10.1002/jbm.b.30128

[55] Guidoin R, ceau D, Rao TJ, King M, Merhi Y, Roy P-E, et al. In vitro and in vivo characterization of an impervious polyester arterial prosthesis: the Gelseal Triaxial® graft. Biomat. 1987;8(6):433–41. DOI: 10.1016/0142-9612(87)90079-2

[56] Marois Y, Chakfé N, Deng X, Marois M, How T, King MW, et al. Carbodiimide cross-linked gelatin: a new coating for porous polyester arterial prostheses. Biomat. 1995;16(15):1131–9. DOI: 10.1016/0142-9612(95)93576-y

[57] Li JK, Wang N, Wu XS. Gelatin nanoencapsulation of protein/peptide drugs using an emulsifier-free emulsion method. J Microencap. 1998;15(2):163–72. DOI: 10.3109/02652049809006846

[58] Cortesi R, Nastruzzi C, Davis SS. Sugar cross-linked gelatin for controlled release: microspheres and disks. Biomat. 1998;19(18):1641–9. DOI: 10.1016/s0142-9612(98)00034-9

[59] Choi YS, Hong SR, Lee YM, Song KW, Park MH, Nam YS. Study on gelatin-containing artificial skin: I. Preparation and characteristics of novel gelatin-alginate sponge. Biomat. 1999;20(5):409–17. DOI: 10.1016/s0142-9612(98)00180-x

[60] Ulubayram K. EGF containing gelatin-based wound dressings. Biomat. 2001;1;22(11): 1345–56. DOI: 10.1016/s0142-9612(00)00287-8

[61] Kim H-W, Yu H-S, Lee H-H. Nanofibrous matrices of poly(lactic acid) and gelatin polymeric blends for the improvement of cellular responses. J Biomed Mater Res. 2008;87A(1):25–32. DOI: 10.1002/jbm.a.31677

[62] Rohanizadeh R, Swain MV, Mason RS. Gelatin sponges (Gelfoam) as a scaffold for osteoblasts. J Mat Sci: Mat Med. 2007;15;19(3):1173–82. DOI: 10.1007/s10856-007-3154-y

[63] Croisier F, Jérôme C. Chitosan-based biomaterials for tissue engineering. Eur Polym J. 2013;49(4):780–92. DOI: 10.1016/j.eurpolymj.2012.12.009

[64] Neves SC, Moreira Teixeira LS, Moroni L, Reis RL, Van Blitterswijk CA, Alves NM, et al. Chitosan/Poly(ɛ-caprolactone) blend scaffolds for cartilage repair. Biomat. 2011;32(4):1068–79. DOI: 10.1016/j.biomaterials.2010.09.073

[65] Stanishevsky AV, Wetuski JD, Yockell-Lelièvre H. Crystallization and stability of electrospun ribbon- and cylinder-shaped tungsten oxide nanofibers. Ceramics Int. 2016;42(1):388–95. DOI: 10.1016/j.ceramint.2015.08.122

[66] Charernsriwilaiwat N, Rojanarata T, Ngawhirunpat T, Sukma M, Opanasopit P. Electrospun chitosan-based nanofiber mats loaded with Garcinia mangostana extracts. Int J Pharm. 2013;452(1-2):333–43. DOI: 10.1016/j.ijpharm.2013.05.012

[67] Kolbuk D, Guimond-Lischer S, Sajkiewicz P, Maniura-Weber K, Fortunato G. The effect of selected electrospinning parameters on molecular structure of polycaprolactone nanofibers. Int J Polym Mater Polym Biomat. 2014;2;64(7):365–77. DOI: 10.1080/00914037.2014.945209

[68] Tao J, Shivkumar S. Molecular weight dependent structural regimes during the electrospinning of PVA. Mater Letters. 2007;61(11-12):2325–8. DOI: 10.1016/j.matlet.2006.09.004

[69] Shenoy SL, Bates WD, Frisch HL, Wnek GE. Role of chain entanglements on fiber formation during electrospinning of polymer solutions: good solvent, non-specific polymer–polymer interaction limit. Polym. 2005;46(10):3372–84. DOI: 10.1016/j.polymer.2005.03.011

[70] Mo X, Xu C, Kotaki M, Ramakrishna S. Electrospun P(LLA-CL) nanofiber: a biomimetic extracellular matrix for smooth muscle cell and endothelial cell proliferation. Biomat. 2004;25(10):1883–90. DOI: 10.1016/j.biomaterials.2003.08.042

[71] Zeng J, Haoqing H, Schaper A, Wendorff JH, Greiner A. Poly-L-lactide nanofibers by electrospinning – Influence of solution viscosity and electrical conductivity on fiber diameter and fiber morphology. e-Polym. 2003;1;3(1). DOI: 10.1515/epoly.2003.3.1.102

[72] Mit-uppatham C, Nithitanakul M, Supaphol P. Ultrafine electrospun polyamide-6 fibers: effect of solution conditions on morphology and average fiber diameter. Macromol Chem Phy. 2004;26;205(17):2327–38. DOI: 10.1002/macp.200400225

[73] Lee KH, Kim HY, Khil MS, Ra YM, Lee DR. Characterization of nano-structured poly(ε-caprolactone) nonwoven mats via electrospinning. Polym. 2003;44(4):1287–94. DOI: 10.1016/s0032-3861(02)00820-0

[74] Koombhongse S, Liu W, Reneker DH. Flat polymer ribbons and other shapes by electrospinning. J Polym Sci, B: Polym Phys. 2001;39(21):2598–606. DOI: 10.1002/polb.10015

[75] Lubasova D, tinova L. Controlled morphology of porous polyvinyl butyral nanofibers. J Nanomat. 2011;2011:1–6. DOI: 10.1155/2011/292516

[76] Stachewicz U, Stone CA, Willis CR, Barber AH. Charge assisted tailoring of chemical functionality at electrospun nanofiber surfaces. J Mater Chem. 2012;22(43):22935. DOI: 10.1039/c2jm33807f

[77] Ding B, Kim H-Y, Lee S-C, Shao C-L, Lee D-R, Park S-J, et al. Preparation and characterization of a nanoscale poly(vinyl alcohol) fiber aggregate produced by an electrospinning method. J Polym Sci, B: Polym Phys. 2002;20;40(13):1261–8. DOI: 10.1002/polb.10191

[78] Reneker DH, Yarin AL, Zussman E, Xu H. Electrospinning of nanofibers from polymer solutions and melts. Adv Appl Mech. 2007;43–346. DOI: 10.1016/s0065-2156(07)41002-x

[79] Greenfeld I, Arinstein A, Fezzaa K, Rafailovich MH, Zussman E. Polymer dynamics in semidilute solution during electrospinning: simple model and experimental observations. Phys Rev E. 2011;84(4). DOI: 10.1103/physreve.84.041806

[80] Lim CT, Tan EPS, Ng SY. Effects of crystalline morphology on the tensile properties of electrospun polymer nanofibers. Appl Phys Lett. 2008;92(14):141908. DOI: 10.1063/1.2857478

[81] Feng B, Duan H, Fu W, Cao Y, Jie Zhang W, Zhang Y. Effect of inhomogeneity of the electrospun fibrous scaffolds of gelatin/polycaprolactone hybrid on cell proliferation. J Biomed Mater Res. 2014;9;103(2):431–8. DOI: 10.1002/jbm.a.35184

[82] Tan EPS, Lim CT. Nanoindentation study of nanofibers. Appl Phys Letters. 2005;87(12): 123106. DOI: 10.1063/1.2051802

[83] Tan EPS, Ng SY, Lim CT. Tensile testing of a single ultrafine polymeric fiber. Biomat. 2005;26(13):1453–6. DOI: 10.1016/j.biomaterials.2004.05.021

[84] Cheng Q, Wang S. A method for testing the elastic modulus of single cellulose fibrils via atomic force microscopy. Compos, A: Appl Sci Manufac. 2008;39(12):1838–43. DOI: 10.1016/j.compositesa.2008.09.007

[85] Gao K, Hu X, Dai C, Yi T. Crystal structures of electrospun PVDF membranes and its separator application for rechargeable lithium metal cells. Mat Sci Eng, B. 2006;131(1-3): 100–5. DOI: 10.1016/j.mseb.2006.03.035

[86] Yang X, Xu Q, Yan N, Sui G, Cai Q, Deng X. Structure and wettability relationship of coelectrospun poly (L-lactic acid)/gelatin composite fibrous mats. Polym Adv Tech. 2010;18;22(12):2222–30. DOI: 10.1002/pat.1749

[87] Wang C, Cheng Y-W, Hsu C-H, Chien H-S, Tsou S-Y. How to manipulate the electrospinning jet with controlled properties to obtain uniform fibers with the smallest diameter?—a brief discussion of solution electrospinning process. J Polym Res. 2010;17;18(1):111–23. DOI: 10.1007/s10965-010-9397-1

[88] Tong H-W, Wang M. Electrospinning of fibrous polymer scaffolds using positive voltage or negative voltage: a comparative study. Biomed Mat. 2010;28;5(5):054110. DOI: 10.1088/1748-6041/5/5/054110

[89] Li D, Ouyang G, McCann JT, Xia Y. Collecting electrospun nanofibers with patterned electrodes. Nano Lett. 2005;5(5):913–6. DOI: 10.1021/nl0504235

[90] Thorvaldsson A, Stenhamre H, Gatenholm P, Walkenström P. Electrospinning of highly porous scaffolds for cartilage regeneration. Biomacromol. 2008;9(3):1044–9. DOI: 10.1021/bm701225a

[91] Fennessey SF, Farris RJ. Fabrication of aligned and molecularly oriented electrospun polyacrylonitrile nanofibers and the mechanical behavior of their twisted yarns. Polym. 2004;45(12):4217–25. DOI: 10.1016/j.polymer.2004.04.001

[92] Casper CL, Stephens JS, Tassi NG, Chase DB, Rabolt JF. Controlling surface morphology of electrospun polystyrene fibers: effect of humidity and molecular weight in the electrospinning process. Macromol. 2004;37(2):573–8. DOI: 10.1021/ma0351975

[93] Lee J, Tae G, Kim YH, Park IS, Kim S-H, Kim SH. The effect of gelatin incorporation into electrospun poly(l-lactide-co-◎-caprolactone) fibers on mechanical properties and cytocompatibility. Biomat. 2008;29(12):1872–9. DOI: 10.1016/j.biomaterials.2007.12.029

[94] Feng B, Tu H, Yuan H, Peng H, Zhang Y. Acetic-acid-mediated miscibility toward electrospinning homogeneous composite nanofibers of GT/PCL. Biomacromol. 2012;10;13(12):3917–25. DOI: 10.1021/bm3009389

[95] Xue J, He M, Liu H, Niu Y, Crawford A, Coates PD, et al. Drug loaded homogeneous electrospun PCL/gelatin hybrid nanofiber structures for anti-infective tissue regeneration membranes. Biomat. 2014;35(34):9395–405. DOI: 10.1016/j.biomaterials.2014.07.060

[96] Chen M, Patra PK, Warner SB, Bhowmick S. Role of fiber diameter in adhesion and proliferation of NIH 3T3 fibroblast on electrospun polycaprolactone scaffolds. Tissue Eng. 2007;13(3):579–87. DOI: 10.1089/ten.2006.0205

[97] Bower DI. An Introduction to Polymer Physics. Cambridge University Press; 2002. p. 343-356. DOI: 10.1017/cbo9780511801280

[98] Gil ES, Frankowski DJ, Bowman MK, Gozen AO, Hudson SM, Spontak RJ. Mixed protein blends composed of gelatin and bombyx m ori silk fibroin: effects of solven-tiInduced crystallization and composition. Biomacromol. 2006;7(3):728–35. DOI: 10.1021/bm050622i

[99] Zhang S. Fabrication of novel biomaterials through molecular self-assembly. Nat Biotech. 2003;21(10):1171–8. DOI: 10.1038/nbt874

[100] Zhang YZ, Feng Y, Huang Z-M, Ramakrishna S, Lim CT. Fabrication of porous electrospun nanofibres. Nanotech. 2006;16;17(3):901–8. DOI: 10.1088/0957-4484/17/3/047

[101] Chew SY, Mi R, Hoke A, Leong KW. The effect of the alignment of electrospun fibrous scaffolds on Schwann cell maturation. Biomat. 2008;29(6):653–61. DOI: 10.1016/j.biomaterials.2007.10.025

[102] Duan B, Dong C, Yuan X, Yao K. Electrospinning of chitosan solutions in acetic acid with poly(ethylene oxide). J Biomat Sci, Polym Ed. 2004;15(6):797–811. DOI: 10.1163/156856204774196171

[103] Zeugolis DI, Khew ST, Yew ESY, Ekaputra AK, Tong YW, Yung L-YL, et al. Electro-spinning of pure collagen nano-fibres – Just an expensive way to make gelatin? Biomat. 2008;29(15):2293–305. DOI: 10.1016/j.biomaterials.2008.02.009

[104] Ghasemi-Mobarakeh L, Prabhakaran MP, Morshed M, Nasr-Esfahani M-H, Ramak-rishna S. Electrospun poly(ɛ-caprolactone)/gelatin nanofibrous scaffolds for nerve tissue engineering. Biomat. 2008;29(34):4532–9. DOI: 10.1016/j.biomaterials.2008.08.007

[105] Sajkiewicz P, Kołbuk D. Electrospinning of gelatin for tissue engineering – molecular conformation as one of the overlooked problems. J Biomat Sci, Polym Ed. 2014;30;25(18):2009–22. DOI: 10.1080/09205063.2014.975392

[106] Geng X, Kwon O, g J. Electrospinning of chitosan dissolved in concentrated acetic acid solution. Biomat. 2005;26(27):5427–32. DOI: 10.1016/j.biomaterials.2005.01.066

[107] De Vrieze S, Westbroek P, Van Camp T, Van Langenhove L. Electrospinning of chitosan nanofibrous structures: feasibility study. J Mat Sci. 2007;15;42(19):8029–34. DOI: 10.1007/s10853-006-1485-6

[108] Van der Schueren L, De Meyer T, Steyaert I, Ceylan Ö, Hemelsoet K, Van Speybroeck V, et al. Polycaprolactone and polycaprolactone/chitosan nanofibres functionalised

with the pH-sensitive dye Nitrazine Yellow. Carbohydrate Polym. 2013;91(1):284–93. DOI: 10.1016/j.carbpol.2012.08.003

[109] Chen F, Li X, Mo X, He C, Wang H, Ikada Y. Electrospun chitosan-P(LLA-CL) nanofibers for biomimetic extracellular matrix. J Biomat Sci, Polym Ed. 2008;19(5):677–91. DOI: 10.1163/156856208784089661

[110] Sangsanoh P, Supaphol P. Stability improvement of electrospun chitosan nanofibrous membranes in neutral or weak basic aqueous solutions. Biomacromol. 2006;7(10): 2710–4. DOI: 10.1021/bm060286l

[111] Elzein T, Nasser-Eddine M, Delaite C, Bistac S, Dumas P. FTIR study of polycaprolactone chain organization at interfaces. J Coll Inter Sci. 2004;273(2):381–7. DOI: 10.1016/ j.jcis.2004.02.001

[112] Pawlak A, Mucha M. Thermogravimetric and FTIR studies of chitosan blends. Thermochim Acta. 2003;396(1-2):153–66. DOI: 10.1016/s0040-6031(02)00523-3

[113] Yang X, Chen X, Wang H. Acceleration of osteogenic differentiation of preosteoblastic cells by chitosan containing nanofibrous scaffolds. Biomacromol. 2009;12;10(10): 2772–8. DOI: 10.1021/bm900623j

[114] Khorshidi S, Solouk A, Mirzadeh H, Mazinani S, Lagaron JM, Sharifi S, et al. A review of key challenges of electrospun scaffolds for tissue-engineering applications. J Tissue Eng Regen Med. 2015;n/a–n/a. DOI: 10.1002/term.1978

[115] Lee JH, g HW, Kang I-K, Lee HB. Cell behaviour on polymer surfaces with different functional groups. Biomat. 1994;15(9):705–11. DOI: 10.1016/0142-9612(94)90169-4

[116] Hong JK, Madihally SV. Three-dimensional scaffold of electrosprayed fibers with large pore size for tissue regeneration. Acta Biomat. 2010;6(12):4734–42. DOI: 10.1016/ j.actbio.2010.07.003

[117] Sangsanoh P, Suwantong O, Neamnark A, Cheepsunthorn P, Pavasant P, Supaphol P. In vitro biocompatibility of electrospun and solvent-cast chitosan substrata towards Schwann, osteoblast, keratinocyte and fibroblast cells. Eur Polym J. 2010;46(3):428–40. DOI: 10.1016/j.eurpolymj.2009.10.029

[118] Hackett JM, Dang TT, Tsai EC, Cao X. Electrospun biocomposite polycaprolactone/ collagen tubes as scaffolds for neural stem cell differentiation. Mater. 2010;19;3(6):3714–28. DOI: 10.3390/ma3063714

[119] Kuppan P, Sethuraman S, Krishnan UM. Interaction of human smooth muscle cells with nanofibrous scaffolds: Effect of fiber orientation on cell adhesion, proliferation, and functional gene expression. J Biomed Mater Res. 2014;5;103(7):2236–50. DOI: 10.1002/jbm.a.35360

[120] Rong D, Chen P, Yang Y, Li Q, Wan W, Fang X, et al. Fabrication of gelatin/PCL electrospun fiber mat with bone powder and the study of its biocompatibility. 2016;4;7(1):6. DOI: 10.3390/jfb7010006

[121] Prabhakaran MP, Venugopal JR, Chyan TT, Hai LB, Chan CK, Lim AY, et al. Electrospun biocomposite nanofibrous scaffolds for neural tissue engineering. Tissue Engg, A. 2008;14(11):1787–97. DOI: 10.1089/ten.tea.2007.0393

[122] Nguyen T-H, Lee B-T. The effect of cross-linking on the microstructure, mechanical properties and biocompatibility of electrospun polycaprolactone–gelatin/PLGA–gelatin/PLGA–chitosan hybrid composite. Sci Technol Adv Mater. 2012;1;13(3):035002. DOI: 10.1088/1468-6996/13/3/035002

[123] Liu H, Hsieh Y-L. Ultrafine fibrous cellulose membranes from electrospinning of cellulose acetate. J Polym Sci, B: Polym Phys 2002;8;40(18):2119–29. DOI: 10.1002/polb.10261

Section 3

State-of-the-Art Polymers

Chapter 5

Electrospinning and Electrospraying Techniques for Designing Antimicrobial Polymeric Biocomposite Mats

Heriberto Rodríguez-Tobías, Graciela Morales and
Daniel Grande

Additional information is available at the end of the chapter

http://dx.doi.org/10.5772/63470

Abstract

Tissue engineering is an interdisciplinary area in which polymeric nanofibers have been exploited either as scaffolds or wound dressings. This chapter overviews the use of versatile electro-hydrodynamic processing techniques for developing antimicrobial nanofibrous mats derived from polymeric biocomposites. The influence of well-known antimicrobial nanoparticles on the physical properties of precursory polymeric solution is discussed, and the consequences of their variations on several important morphological parameters, namely average fiber diameter and orientation, porosity, pore size, among others, are considered. Moreover, other specific properties of composite fibers conferred by the nanoparticles are reviewed as well as their use toward the design of multifunctional polymeric mats.

Keywords: antimicrobial nanoparticles, biocomposites, electrospinning, electrospraying, polymeric nanofibers

1. Introduction

Electro-hydrodynamic techniques, i.e., electrospinning and electrospraying, have been extensively explored in the biomedical area for the last 15 years, probably due to the submicrometric diameter exhibited by the obtained fibers; and consequently, to the large surface area to volume ratio and high porosity with interconnected voids formed between the fiber structures. These morphological features and size scale are suitable to mimic the natural extracellular matrix (ECM), thus promoting cell attachment and proliferation. Therefore, fibrous materials have been catalogued as excellent materials for tissue engineering, and a vast number of publications related to their applications as wound dressings and/or cell-based scaffolds have been recorded [1, 2].

Electrospinning and electrospraying techniques are based on identical principles and the equipment typically consists of a syringe containing the solution that passes through a capillary (needle) and the flow rate is controlled by a pump. In turn, a power source applies several tens of kV on the needle, and therefore on the polymer solution. The applied high voltage causes a significant density of charges on the droplet which protrudes into the tip of the needle up to a point where the repulsive forces of the charges exceed the surface tension and polymer jet or drop is eventually produced and then captured in the grounded collector, and the solvent being evaporated during the passage from the tip of the needle to the collector. The final morphology of the obtained polymeric material depends on the physical properties of the polymer solution as well as the electro-hydrodynamic device parameters [3, 4]. The effects of these parameters on the fiber diameters and morphologies are gathered in **Table 1**. When the polymer solution concentration is enough to promote chain entanglements, a polymer jet is formed. On the contrary, if the polymer solution concentration is too low, chain entanglements will not occur and droplets will be sprayed from the needle, and the electrospraying process takes place [5–7].

Parameter	Effect on fiber morphology
Increase in applied voltage	Fiber diameter initially decreases, then increases (not monotonic)
Increase in flow rate	Fiber diameter increases (beaded morphologies occur if the flow rate is to high)
Increase in needle-to-collector distance	Fiber diameter decreases (beaded morphologies occur if the needle-to-collector distance is too short)
Increase in polymer concentration (viscosity)	Fiber diameter increases (within optimal range)
Increase in solution conductivity	Fiber diameter decreases (broad diameter distribution)
Increase in solvent volatility	Fibers exhibit surface microtexture (pores on their surfaces, which may increase surface area)

Source: Ref. [3].

Table 1. Effect of different parameters on the morphology of polymeric fibers derived from electrospinning process.

Currently, there is a trend to develop nanocomposite fibers by means of the incorporation of nanoparticles in a polymer solution; the resulting materials consist of nanoparticles-embedded polymeric fibers with enhanced or new properties. On the other hand, an innovative approach has been recently employed in order to obtain nanoparticles-coated polymeric fibers that involve a tandem process, i.e., simultaneous electrospinning of a polymer solution and the electrospraying of a nanoparticles dispersion, whose configuration is illustrated in **Figure 1**. This combination of electro-hydrodynamic techniques gives rise to a higher extent

http://dx.doi.org/10.5772/63470

of exposed nanoparticles, thus increasing the surface-dependent properties of nanocomposite fibers, such as antimicrobial, optical, and catalytic properties, among others [5–10].

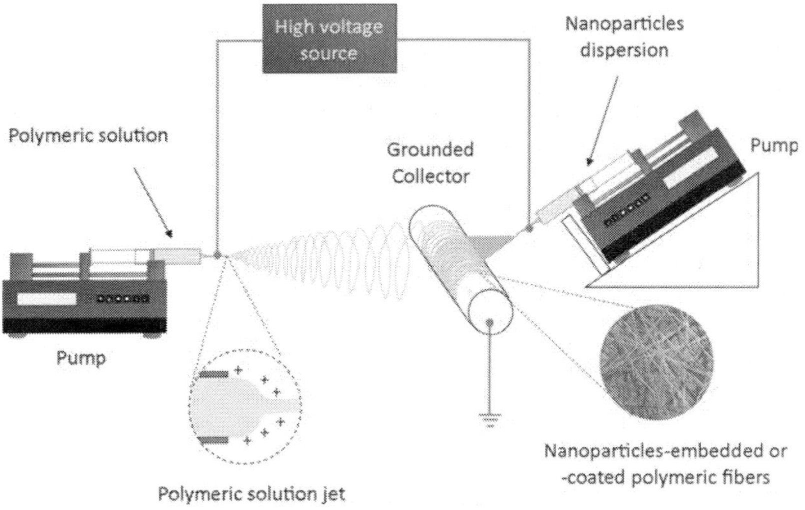

Figure 1. Typical configuration of an electrospinning/electrospraying tandem equipment.

2. Antimicrobial biopolymeric nanofibers by electrospinning and electrospraying

Materials designed by electrospinning and electrospraying techniques are predominantly used as mats in tissue engineering. The mats derived from electro-hydrodynamic techniques indeed possess morphology similar to the natural ECM that is composed of a submicron fiber network of proteins and glycosaminoglycans. Electrospun nanofibrous scaffolds may act as a support for cell adhesion and subsequent proliferation, thus providing cells with the appropriate microenvironment of chemical and physical cues, including cell-matrix and cell-cell interactions, and eventually forming tissues with suitable mechanical and biological properties [3, 4].

Regarding the polymeric matrix, biocompatible, biodegradable, and/or bioabsorbable polymers are desirable, since the mats gradually degrade leaving only the new tissue and also avoid a subsequent surgery [2–4]. Several aliphatic polyesters possess the aforementioned features, among which poly(ε-caprolactone), poly(3-hydroxybutyrate), poly(4-hydroxybutyrate), the corresponding copolymers with 3- and 4-hydroxyvalerate, poly(lactic acid), and the respective copolymer with glycolic acid could be cited. Some polysaccharides have also been used, chitosan being the most noticeable.

However, polymeric mats derived from electro-hydrodynamic techniques entail a huge inconvenient, their fibrous/porous morphology is also prone to the adhesion of pathogenic microorganisms, thus facilitating the occurrence of serious infections [11–13]. Several strategies have been considered in order to overcome one such drawback, the incorporation of organic compounds being the most prominent, and more recently, nanoparticles with well-recognized antimicrobial properties. Polymeric nanofibers containing antimicrobial nanoparticles can exhibit several advantages compared to typical organic compound-loaded polymeric analogues, such as higher thermal stability, enhanced mechanical performance, or biocompatibility, depending on the chemical nature of nanoparticles [14–17]. In this context, the design of antimicrobial polymeric fibers containing different nanoparticles will be addressed in the next subsections.

2.1. Polymeric mats containing silver nanoparticles as antimicrobial additives

Silver nanoparticles (nano-Ag) have extensively been investigated as antimicrobial agents against a broad variety of microorganisms, from bacteria to viruses [18]. Silver in its metallic state is inert but it reacts with the moisture in the skin and the fluid of the wound and gets ionized. The ionized silver is highly reactive, as it binds to tissue proteins and brings structural changes in the bacterial cell wall and nuclear membrane leading to cell distortion and death. Silver also binds to bacterial DNA and RNA by denaturing and inhibits bacterial replication. The antimicrobial performance of silver is related to the amount of silver and the rate of silver ions released [19]. Although silver nanoparticles have been used as cream or gel, there is a trend to incorporate them into polymeric wound dressings in order to prevent infections, and in turn, promote wound healing [20–22].

Poly(vinyl alcohol) (PVA) is a biocompatible and nontoxic polymer suitable for the design of antibacterial wound dressings derived from electro-hydrodynamic techniques. Additionally, PVA is a water soluble polymer, thus making its fibers elaboration an ecofriendly and innocuous process. Hong [23] engineered PVA-based mats containing nano-Ag by means of precursory solution of the corresponding polymer and $AgNO_3$ with a subsequent thermal or ultraviolet (UV) posttreatment. The obtained PVA-based mats exhibited excellent antibacterial activity against *Staphylococcus aureus* and *Klebsiella pneumoniae*. Although the water solubility of PVA might be a drawback when using it as wound dressing, this can be overcome by simple thermal treatment without negative effect on final properties.

Likewise, PVA solutions have also been mixed with nano-Ag and montmorillonite (MMT) in order to obtain the corresponding fibers with antibacterial properties and enhanced thermal stability (conferred by MMT). At a fixed MMT concentration, an increase in nano-Ag from 1 to 5 wt% led to higher average fiber diameters, which was attributed to an increase in solution viscosity, thus illustrating that nanoparticles seriously affected the final morphology of mats. Nano-Ag triggered the inhibition growth of both Staphylococcus *S. aureus* and *Escherichia coli* [24].

In other studies, nano-Ag was embedded into electrospun nanofibers based on chitosan. This polysaccharide has attracted the industrial and academic attention, as it is a bio-sourced material and possesses inherent antimicrobial activity. Nonetheless, electrospinning of neat chitosan is complicated due to its limited solubility; therefore, it has typically been mixed with other biocompatible polymers such as poly(ethylene oxide) (PEO) and PVA. An et al. [25] designed antibacterial chitosan/PEO nanofiber by means of incorporation of nano-Ag. The results led to the conclusion that viscosity and surface tension of the precursory solution were not affected by the nano-Ag concentration, but the electrical conductivity was proportionally increased by the augmentation of the metallic nanoparticles. The authors argued that residual Ag^+ ions were responsible for this behavior. Another important fact was that the presence of nano-Ag promoted the formation of fibers with smooth surface and the average fiber diameter was decreased, as expected by the increase in electrical conductivity. Furthermore, the obtained polymeric mats with embedded nano-Ag (2.2 wt%) exhibited higher tensile strength than that for neat mats, while the corresponding elongation was reduced. Regarding the antibacterial activity against *E. coli*, the chitosan/PEO-based mats significantly decreased the proliferation of this bacterium in a short period of time (6–8 h). Penchev et al. [26] have developed a one step-procedure for the preparation of cross-linked chitosan/PEO/Ag mats whose stability in aqueous environment render these new materials promising candidates for applications, such as wound-healing dressings or antibacterial filters.

Other biodegradable polyesters, such as poly(lactic acid) and poly(ε-caprolactone) have been used as matrices for the elaboration of nanofibers with Ag nanoparticles. Kim et al. [21] investigated the influence of $AgNO_3$ concentration on the viscosity, electrical conductivity, and surface tension of precursory polymeric solution, and a relationship with the final morphology was proposed. It is important to mention that after electrospinning process, $AgNO_3$ was reduced to Ag. The incorporation up to 9 wt% of $AgNO_3$ did not alter the viscosity or surface tension, but the electrical conductivity was considerably increased, as observed in other polymeric systems. Regarding the morphology, the increase in $AgNO_3$ concentration (i.e., electrical conductivity) from 3 to 6 wt% led to thinner PLA fibers, interestingly, an optimal $AgNO_3$ concentration was perceived since thicker fibers were formed at 9 wt% $AgNO_3$. The mats containing nano-Ag showed growth inhibition values higher than 99%, regardless of the type of bacterium and nanoparticles concentration. Xu et al. [27] conducted a similar investigation but with higher content of nano-Ag from 9 to 32 wt%. The author demonstrated that at high nano-Ag concentration, average fiber diameter increased probably due to interactions with PLA, and consequently increased the solution viscosity. The antibacterial activity against *E. coli* and *S. aureus* was practically identical to that for PLA mats with lower nano-Ag concentration as that reported by Kim.

Recently, Augustine et al. [28] prepared PCL fibers containing embedded silver nanoparticles. Regarding the morphological features, the fiber diameter was decreased by the presence of nanoparticles. The authors claimed that nanoparticles increased the electrical conductivity; however, no evidence was given. EDX analysis showed that the higher the silver content, the more significant the formation of aggregates, and the latter were exposed on the surface of the PCL fibers. The aggregation of silver nanoparticles had an impact on the mechanical properties of the fibrous materials, since at relatively low concentrations (0.05 and 0.5 wt%) both; elongation and fracture resistance were enhanced, while incorporating 1 wt%, the mechanical properties had no changes compared with blank PCL fibers. Concerning the performance of the fibers as antibacterial materials, disk diffusion tests proved that the growth inhibition area of *S. aureus* is doubled at a nanoparticles concentration as low as 0.5 wt %, while the bacterium *E. coli* showed greater resistance.

Actually, there is a limited investigation related to the use of mere electrospraying or its combination with electrospinning for obtaining antimicrobial nanocomposites mats with Ag nanoparticles. Park et al. [29] successfully obtained PVA/montmorillonite/silver hybrid particles by means of an aqueous solution of the corresponding polymer and inorganic nanoparticles subjected to electrospraying. The composite polymeric particles were formed only at low PVA molar mass and concentration, and the transmission electron microscopy (TEM) images evidenced the exfoliated montmorillonite and well-dispersed Ag nanoparticles, regardless of their concentration in the precursory solution. The hybrid polymeric particles containing the combination of nanostructures possess a higher thermal stability attributed to montmorillonite, and no interference of Ag was perceived. Electrosprayed materials having Ag nanoparticles inhibited the proliferation of the *S. aureus* strain, while those without the mentioned metallic nanoparticles failed, thus showing the potential of electrospraying technique for obtaining polymeric nanoparticles with antibacterial activity.

2.2. Antimicrobial polymeric mats containing metal oxides nanoparticles

Metal oxides nanoparticles are another type of particles not only used as antimicrobial agents but also they can also impart other interesting optical and electrical properties, which can potentiate their use as a semiconductors, electroluminescent devices, thermoelectric materials, and for environmental decontamination applications [30–32]. Among the main metal oxide nanoparticles used as antimicrobial agents, zinc oxide (ZnO) and titanium dioxide (TiO_2) can be mentioned. They are addressed in the following subsections.

In this context, TiO_2 has also been embedded in biopolymer fibers to confer antimicrobial properties and its potential use in tissue regeneration. Lee et al. [33] developed multifunctional PVA fibers with excellent growth inhibition against *S. aureus* and *K. pneumoniae*. Furthermore, the obtained fibrous materials functioned as UV filter membranes and degrading agents of ammonia and formaldehyde. These features were accomplished with relatively low nanoparticles content (2–3 wt% respect to the polymer). It is noteworthy that the effect of the presence and/or concentration of the nanoparticles on the precursory solution properties was not studied.

Son et al. [34] have also exploited the antibacterial activity of TiO_2 to prepare polymeric fibers based on chitosan/PVA blends. The experimental results demonstrated a direct relationship between the content of chitosan and the viscosity of the precursory solution. The authors also found that at chitosan concentration greater than 7 wt% (relative to volume of solution), it was not possible to obtain uniform morphology, i.e., fibers with numerous ovoid defects were obtained. Interestingly, the presence of TiO_2 nanoparticles did not change the morphological features, namely average fiber diameter. Authors also established that the diameter of the fibers (270–360 nm) had no significant effect on the antibacterial activity against *E. coli* and *S. aureus*. Furthermore, good growth inhibition (approximately 83–85%) was obtained with TiO_2 concentrations as low as 0.04 wt% into the polymeric blend. When comparing these results with those for fibers containing silver nitrate, it was evidenced that the presence of the latter was more effective for growth inhibition of both bacteria.

Rashkov's group has thoroughly studied the combination of electro-hydrodynamic techniques. In 2013, this group prepared submicrometric fibers based on PHB and TiO_2 nanoparticles by mere electrospinning and electrospinning/electrospraying tandem technique [35]. The fibers obtained by mere electrospinning of PHB/TiO_2 presented defects along them, which were attributed to the formation of agglomerates, and consequently, disturbance of the solution flow. Furthermore, the nanoparticles were successfully deposited on PHB fibers by the tandem process, although some aggregates were perceived with up to 1 μm in size. In order to overcome the aggregation of nanoparticles, the authors added chitosan oligomers to the precursory solution, keeping in mind that it might act as a dispersant. However, the selected approach was not favorable, since aggregates as big as 10 μm were detected, which was attributed to the increase in the dispersion viscosity, thus forming large droplets during the electrospraying process. The authors also implemented an impregnation process in order to obtain the TiO_2-covered PHB fibers, whose morphological results were not encouraging due to the formation of large aggregates. The materials obtained showed excellent pigment degradation of pollutants; methylene blue (used as pigment model) was entirely degraded after only 3 h in the presence of the electrospun materials. The composite fibers obtained by the different strategies showed significant antibacterial activity against *E. coli*, with the advantage of allowing for the adhesion and development of human mesenchymal stem cells, which was attributed to the morphological characteristics suitable for applications in the field of tissue engineering.

Another contribution from Rashkov's group was also related to the combination of electro-hydrodynamic techniques, by which a series of PHB-based fibers with TiO_2 and iron oxide (Fe_3O_4) nanoparticles were designed in order to obtain multifunctional mats [36]. The authors carried out the electrospinning of PHB containing TiO_2 and Fe_3O_4, the combination of electrospinning of PHB and electrospraying of Fe_3O_4 or Fe_3O_4/TiO_2 dispersion, and electrospinning of PHB/Fe_3O_4 combined with electrospraying of TiO_2. In order to stabilize the nanoparticles dispersions, low amounts of chitosan or its oligomers in acetic acid were used (0.5–2%, w/v). Regarding the morphology, the composite PHB fibers derived from electrospinning exhibited a significant number of beads, while the materials produced by the combined techniques of electrospinning/electrospraying showed aggregates on the surface. Furthermore, a smaller amount of nanoparticles than initial feed was determined by X-ray photoelectron spectroscopy (XPS), which was ascribed to the formation of a dense layer of the stabilizing agent used (chitosan or oligomers). Since the antibacterial activity of identical materials had already explored, the new series of electrospun PHB fibrous mats were used as membranes for the degradation of pollutants pigments. The degradation activity was higher for the fibrous materials obtained by the conjunction of electro-hydrodynamic techniques, due to higher surface area exposed to the solution containing the pigment. It is noteworthy that no information of the effect of metal oxide nanoparticles on viscosity or electrical conductivity was reported in this work.

Zinc oxide (ZnO) nanoparticles are considered as antimicrobial agents that can strongly compete with silver nanoparticles, especially for their simple and inexpensive synthesis as well as its effectiveness in eliminating several pathogen microorganisms [37–40]. ZnO nanoparticles have been exploited for the development of fibrous materials based on biodegradable polymers, and some of these reports shall be addressed next.

Sodium alginate is a polyelectrolyte that has excellent biocompatibility, biodegradability, and ease of dissolution in water; however, it is not suitable for the preparation of fibers by electrospinning. In order to facilitate the fiber formation, Shalumon et al. [41] mixed sodium alginate with different proportions of PVA, which lowers the inherent brittleness of the sodium alginate, and the authors used ZnO nanoparticles (diameter = 100 nm) as an antibacterial agent. The resulting fibers were crosslinked using glutaraldehyde to increase their dimensional stability. Since sodium alginate is a polyelectrolyte, this increased electrical conductivity of the solution. This behavior was also observed when the nanoparticles were incorporated; however, the effect was not so marked but until a nanoparticles concentration higher than 5 wt%. Similarly, the viscosity of the solution was increased, due to the presence of alginate or nanoparticles, and this parameter had the most important effect on fiber morphology, specifically, the fiber diameter which slightly increased. The authors also demonstrated that the presence of the alginate promoted higher thermal stability of the fibers, while the nano-metal oxide did not contribute significantly to the retard in thermal degradation. Concerning the antibacterial properties, the fiber materials showed excellent antibacterial activity against *E. coli* and *S. aureus*, which was proportional to the content of nanoparticles. Conversely, the adhesion of mice fibroblasts was diminished as the concentration of ZnO was increased.

Augustine et al. [42–44] have intensively investigated the influence of ZnO nanoparticles on the morphology, bactericidal performance, and cell proliferation on mats based on PCL fibers. In 2014, these authors reported that the incorporation of a relatively low amount of nano-ZnO (0.1–0.9 wt%) did not significantly affect the morphology of PCL fibers. Conversely, when a concentration of 1 wt% is exceeded, the morphology undergoes significant changes mainly in the appearance of pores and the fiber diameters, which they attributed to the increased viscosity of the precursor solution and the charge density; however, electrical conductivity or rheological analyses were not presented. EDX spectroscopic studies showed that by increasing the concentration of ZnO nanoparticles, they can be exposed on the surface of the PCL fibers, while the existence of hydrogen bond interactions between hydroxyl groups on nanoparticles surface and the carbonyl groups of the PCL was established by Fourier transform infrared (FTIR). Regarding antibacterial activity, it was determined that a lower concentration of 5 wt% was not enough to inhibit the proliferation of *E. coli* and *S. aureus*. Interestingly, PCL/ZnO fibrous mats did not show cytotoxicity against adult goat fibroblasts (*in vitro* study), which makes them promising as antibacterial biodegradable scaffolds [43]. Concurrently, Augustine et al. showed that the PCL/ZnO composite fibers had the ability to heal wounds on the skin of animals without showing inflammation of tissues when ZnO nanoparticles was below 4 wt%, which allows for further use as wound dressings and/or scaffolds with bactericidal properties [44].

Poly(L-lactide) has also been subjected to both the electrospinning process and the electrospinning/electrospraying tandem technique. Rashkov's group [45] designed fibrous mats whose inorganic phase consisted of ZnO nanoparticles. The authors took advantage of the photocatalytic and antibacterial properties of ZnO to generate purifying membranes. The fibers obtained by mere electrospinning showed high efficiency to degrade methylene blue and an azo pigment (95% and 65% degradation, respectively) in only 300 min. Meanwhile, the fibers resulting from electrospinning/electrospraying degraded both pigments in a shorter time (approximately 180 min). This behavior was explained in terms of a larger nanoparticles surface area prone to degrade the pigments. Furthermore, the materials had a considerable efficiency for killing *S. aureus*, in a greater extent for the PLA/ZnO fibers derived from the combined technique. Considering the biocompatibility, the content of nanoparticles was significantly high (greater than 45 wt%), which is not beneficial when trying to use such materials as surgical implants, since it has been shown that ZnO nanoparticles contents higher than 5 wt% is not suitable for adhesion and proliferation of tissue cells.

In order to get further insight into the antibacterial activity of PLA-based fibers containing ZnO nanoparticles, our research groups in Mexico and France have developed electrospun and electrospun/electrosprayed mats derived from poly(D,L-lactide) and low amounts of ZnO nanoparticles [10]. Scanning electron microscopy (SEM) images of the obtained mats are disclosed in **Figure 2**. It was noted that the incorporation of nanoparticles provoked slight variations in viscosity and conductivity of precursory solutions, and no significant effect on the morphological parameters (namely, average fiber diameter, pore size, and porosity) of electrospun PLA-based mats was observed. Furthermore, the presence of nano-ZnO enhanced the mechanical performance of these materials, exhibiting an optimal concentration of nanoparticles for 3 wt%, probably due to hydrogen bonds between hydroxyl groups on ZnO nanoparticles and carbonyl groups of PLA. Regarding the antibacterial properties of mats, they were shown to be dependent on the type of bacteria, being *E. coli* less sensitive to the presence of nanoparticles. On the other hand, when electrospinning/electrospraying was used with at least 1 wt% of ZnO nanoparticles, values higher than 94% of growth inhibition of *S. aureus* were achieved, while simple electrospun mats did not inhibit the bacterium growth for the same ZnO concentration. The results obtained make ZnO-coated fibrous PLA mats potential candidates for applications related to wound dressing materials.

A similar protocol used for PLA/ZnO mats was adapted for developing PHB-based antibacterial fibers [46]. The analysis of the physical properties (viscosity and electrical conductivity) showed that the presence of ZnO nanoparticles had an insignificant effect at low concentration (1 and 3 wt%) and a slight effect at ZnO content of 5 wt%. Consequently, the final morphology of corresponding nanofibers was not altered. The obtained PHB/nano-ZnO mats showed uniform fiber morphology with an average porosity *ca.* 85% with enhanced thermal stability compared to that of pristine PHB. Differential scanning calorimetry was also used to determine the influence of ZnO nanoparticles on the phase transitions of as-spun PHB nanofibers; it was thus shown that the nanoparticles promoted the formation of different

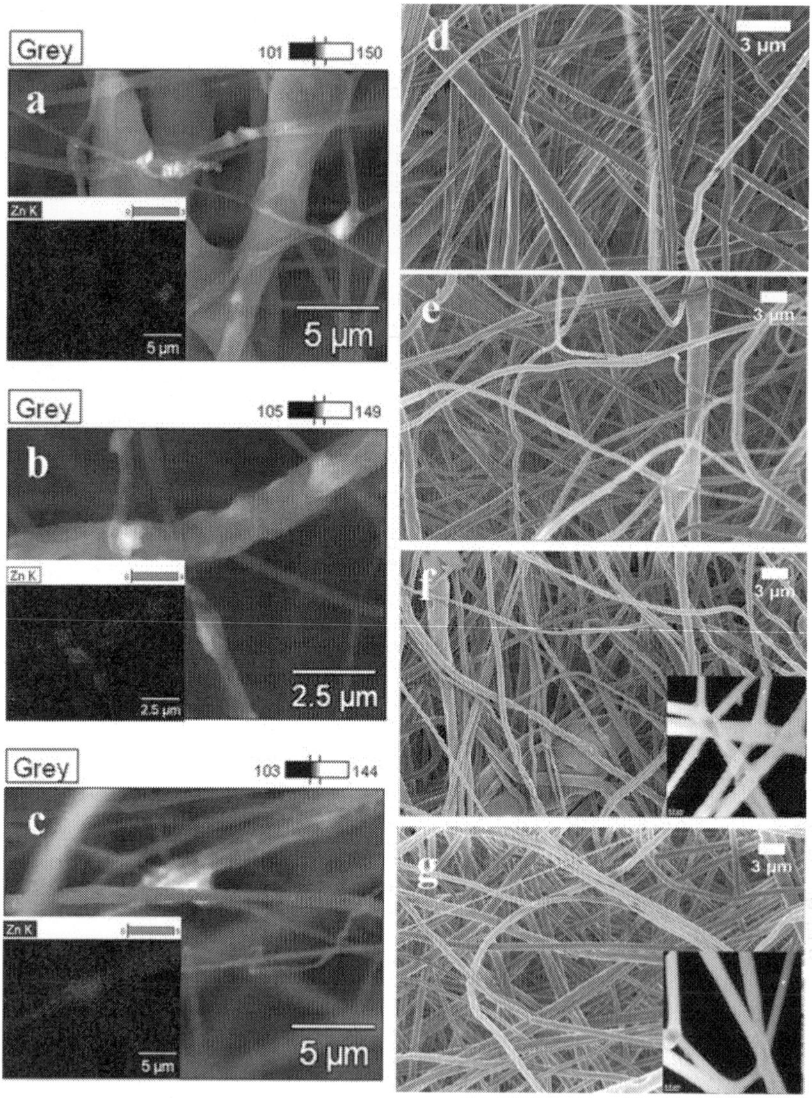

Figure 2. SEM images of PLA/ZnO mats obtained by means of electrospinning/electrospraying with (a) 1, (b) 3, and (c) 5 wt% ZnO, respectively, and simple electrospinning with (d) 0, (e) 1, (f) 3, and (g) 5 wt% ZnO, respectively. Insets: (a)-(c) EDX maps;(f), (g) TEM images.

crystalline entities. Furthermore, the antibacterial performance against *E. coli* and *S. aureus* was proved to be dependent on the elaboration technique, being higher for electrospinning/

http://dx.doi.org/10.5772/63470

electrospraying tandem technique, thus permitting the design of novel bacteriostatic or bactericidal PHB/nano-ZnO nanofibrous composites.

Copper oxide nanoparticles are other well-recognized antibacterial agents that have been incorporated into polymeric matrices, but not specifically biodegradable. The studies related to copper nanoparticles used as antimicrobial agent in electrospun polymers are limited, probably due to the instability of this metal oxide to environmental conditions, which can change the oxidation state of copper, thus changing its final properties.

However, Ungur and Hrůza [47] used a semi-industrial scale electrospinning (roller spinning method) for developing poly(urethane) (PU) nanofibers modified with copper oxide (II) (CuO). The roller spinning method consists of an aluminum rotating cylinder body with spikes which was partially immersed into the polymer solution. High voltage is connected to the rotating roller. As the solvent evaporates, the jets of polymer solution are transformed and the solid nanofibers are obtained before reaching the collector electrode. The nanofibers were collected on polypropylene (PP) spun bond nonwoven antistatic material. Regarding the physical properties of PU solutions, the authors demonstrated that low CuO content (2.5 and 5 wt%) have no noticeable influence on the rheological behavior of solutions within time at low shear rate. However, the viscosity of PU solutions was increased in the presence of higher concentrations of antimicrobial agent (7, 9.5, 12, and 15 wt%). On the other hand, electrical conductivity was identical for all solutions, i.e., the presence of CuO did not affect this physical property. Interestingly, despite the increase in solution viscosity by the incorporation of CuO, average fiber diameter of pristine PU and PU/CuO mats exhibited a slight variation ranging from 250 to 300 nm. The result of EDX analysis demonstrated the significant decrease in the amount of Cu on the nanofiber surface for the fibers with 15% CuO, probably due to their aggregation within the fiber structure. The obtained materials demonstrated high antibacterial activity against Gram-negative bacterial strain *E. coli* and Gram-positive strain *S. aureus* at relatively low CuO contents (2.5 wt%) and it was also demonstrated that the incorporation of higher amount of nanoparticles did not improve the antibacterial properties.

Poly(acrylonitrile) (PAN) possesses good mechanical and thermal properties; therefore, it is a suitable polymer for designing purifying filters and (no biodegradable) wound dressings by electro-hydrodynamic techniques. For instance, Zhang et al. [48] employed a method that allowed for the synthesis of zero-valent copper (Cu^0) nanoparticles in/on PAN nanofibers by two steps, including the preparation of polyacrylonitrile/copper(II) (PAN/Cu^{2+}) nanofibers by electrospinning, and the preparation of PAN/Cu^0 nanofibers by high-pressure hydrogenation reduction. Regarding the morphology of PAN/Cu^{2+} nanofibers, the average diameter ranged from 200 to 600 nm, and the nanofibers were becoming more and thicker with the increase in copper nitrate. The authors also demonstrated that the use of high-pressure hydrogenation reduction chamber did not change the fiber morphology, and a good dispersion of nanoparticles was preserved, showing Cu^0 aggregates lower than 20 nm. The PAN/Cu^0 showed bacteriostatic effect against *S. aureus* (i.e., the survival bacteria population was lower than those counted in pristine polymer, but higher than the bacteria population inoculated), thus showing the potential use of these fibers as antifouling materials.

3. Conclusions and perspectives

Electro-hydrodynamic techniques, namely electrospraying and electrospinning, are powerful approaches for developing material with morphological features suitable for tissue engineering applications. The incorporation of nanoparticles allows for the generation of multifunctional mats that can be used in different applications ranging from filters for pollutant removal to wound dressings. Regarding the pollutant removal, metal oxides (mainly ZnO and TiO_2) are the more promising nanoparticles, while for tissue engineering a further insight into the biocompatibility should be obtained. Comparing the electro-hydrodynamic techniques, the electrospraying process enables higher nanoparticles surface area prone to act as antibacterial agents or degrading materials, regardless of the presence of aggregates obtained in most of the cases. Finally, certain nanoparticles promote other specific properties, such as thermal stability, original crystallization behavior, enhanced mechanical properties, etc. which can lead to the diversification of polymeric/nanoparticles-based mats.

Electro-hydrodynamic techniques have a great potential in many applications; however, there is a need for developing novel or adapted equipment that enables the fabrication of polymeric mats in a higher scale.

Acknowledgements

The authors thank the National Council of Science and Technology (CONACyT, Mexico) for providing H. Rodríguez-Tobías with a Ph.D. grant. They also thank the Ministry of Foreign Affairs and the Ministry of National Education and Higher Education and Scientific Research from France and CONACYT from Mexico for financial support of the research stay in the ICMPE (Thiais, France).

Author details

Heriberto Rodríguez-Tobías[1], Graciela Morales[2*] and Daniel Grande[2*]

*Address all correspondence to: graciela.morales@ciqa.edu.mx

*Address all correspondence to: grande@icmpe.cnrs.fr

1 Polymer Synthesis Department, Centro de Investigación en Química Aplicada, Saltillo, Mexico

2 Complex Polymer Systems Laboratory, Institut de Chimie et des Matériaux Paris-Est, UMR 7182 CNRS-Université Paris-Est Créteil, Thiais, France

References

[1] Bhardwaj N, Kundu SC. Electrospinning: a fascinating fiber fabrication technique. Biotechnol Adv. 2010;28(3):325–347. DOI: 10.1016/j.biotechadv.2010.01.004.

[2] Zahedi P, Rezaeian I, Ranaei-Siadat S-O, Jafari S-H, Supaphol P. A review on wound dressings with an emphasis on electrospun nanofibrous polymeric bandages. Polym Adv Technol. 2009;21(2):77–95. DOI: 10.1002/pat.1625.

[3] Sill TJ, von Recum HA. Electrospinning: applications in drug delivery and tissue engineering. Biomaterials. 2008;29(13):1989–2006. DOI: 10.1016/j.biomaterials. 2008.01.011.

[4] Villarreal-Gómez LJ, Cornejo-Bravo JM, Vera-Graziano R, Grande D. Electrospinning as a powerful technique for biomedical applications: a critically selected survey. J Biomater Sci Polym Ed. 2016;27(2):157–176. DOI: 10.1080/09205063.2015.1116885.

[5] Jaworek A, Krupa A, Lackowski M, Sobczyk AT, Czech T, Ramakrishna S, Sundarrajan S, Pliszka D. Nanocomposite fabric formation by electrospinning and electrospraying technologies. J Electrost. 2009;67(2–3):435–438. DOI: 10.1016/j.elstat.2008.12.019.

[6] Bock N, Woodruff MA, Hutmacher DW, Dargaville TR. Electrospraying, a reproducible method for production of polymeric microspheres for biomedical applications. Polymers. 2011;3(4):131–149. DOI: 10.3390/polym3010131.

[7] Vitchuli N, Shi Q, Nowak J, Kay K, Caldwell JM, Breidt F, Bourham M, McCord M, Zhang X. Multifunctional ZnO/Nylon 6 nanofiber mats by an electrospinning–electrospraying hybrid process for use in protective applications. Sci Tech Adv Mater. 2011;12(5):055004. DOI: 10.1088/1468-6996/12/5/055004.

[8] Ramier J, Bouderlique T, Stoilova O, Manolova N, Rashkov I, Langlois V, Renard E, Albanese P, Grande D. Biocomposite scaffolds based on electrospun poly(3-hydroxybutyrate) nanofibers and electrosprayed hydroxyapatite nanoparticles for bone tissue engineering applications. Mater Sci Eng C. 2014;38:161–169. DOI: 10.1016/j.msec. 2014.01.046.

[9] Virovska D, Paneva D, Manolova N, Rashkov I, Karashanova D. Electrospinning/electrospraying vs. electrospinning: a comparative study on the design of poly(l-lactide)/zinc oxide non-woven textile. Appl Surf Sci. 2014;311:842–850. DOI: 10.1016/j.apsusc.2014.05.192.

[10] Rodríguez-Tobías H, Morales G, Ledezma A, Romero J, Grande D. Novel antibacterial electrospun mats based on poly(d,l-lactide) nanofibers and zinc oxide nanoparticles. J Mater Sci. 2014;49(24):8373–8385. DOI: 10.1007/s10853-014-8547-y.

[11] Speranza G, Gottardi G, Pederzolli C, Lunelli L, Canteri R, Pasquardini L, Carli E, Lui A, Maniglio D, Brugnara M, Anderle M. Role of chemical interactions in bacterial

adhesion to polymers surfaces. Biomaterials. 2004;25:2029–2037. DOI: 10.1016/j.biomaterials.2003.08.061.

[12] Pavithra D, Doble M. Biofilm formation, bacterial adhesion and host response on polymeric implants: issues and prevention. Biomed Mater. 2008;3(3):034003. DOI: 10.1088/1748-6041/3/3/034003.

[13] Armentano I, Arciola CR, Fortunati E, Ferrari D, Mattioli S, Amoroso CF, Rizzo J, Kenny JM, Imbriani M, Visai L. The interaction of bacteria with engineered nanostructured polymeric materials: a review. Scientific World J. 2014;2014:410423. DOI: 10.1016/j.polymdegradstab.2010.06.007.

[14] Schiffman JD, Elimelech M. Antibacterial activity of electrospun polymer mats with incorporated narrow diameter single-walled carbon nanotubes. ACS Appl Mater Interfaces. 2011;3(2):462–468. DOI: 10.1021/am101043y.

[15] Xu X, Yang Q, Wang Y, Yu H, Chen X, Jing X. Biodegradable electrospun poly(l-lactide) fibers containing antibacterial silver nanoparticles. Eur Polym J. 2006;42(9):2081–2087. DOI: 10.1016/j.eurpolymj.2006.03.032.

[16] Shalumon KT, Anulekha KH, Nair S V, Chennazhi KP, Jayakumar R. Sodium alginate/poly(vinyl alcohol)/nano ZnO composite nanofibers for antibacterial wound dressings. Int J Biol Macromol. 2011;49(3):247–254. DOI: 10.1016/j.ijbiomac.2011.04.005.

[17] Gao Y, Truong YB, Zhu Y, Kyratzis IL. Electrospun antibacterial nanofibers: production, activity, and in vivo applications. J Appl Polym Sci. 2014;131:40797. DOI: 10.1002/app.40797.

[18] Rai M, Yadav A, Gade A. Silver nanoparticles as a new generation of antimicrobials. Biotechnol Ad. 2009;27:76–83. DOI: 10.1016/j.biotechadv.2008.09.002.

[19] Lansdown AB. Silver I: its antibacterial properties and mechanism of action. J Wound Care. 2002;11(4):125–130. DOI: http://dx.doi.org/10.12968/jowc.2002.11.4.26389.

[20] Hong KH, Park JL, Sul IH, Youk JH, Kang TJ. Preparation of antimicrobial poly(vinyl alcohol) nanofibers containing silver nanoparticles. J PolymSci Phys. 2006;44(17):2468–2474. DOI: 10.1002/polb.20913.

[21] Kim ES, Kim SH, Lee CH. Electrospinning of polylactide fibers containing silver nanoparticles. Macromol Res. 2010;18(3):215–221. DOI: 10.1007/s13233-010-0316-4.

[22] Rujitanaroj P, Pimpha N, Supaphol P. Wound-dressing materials with antibacterial activity from electrospun gelatin fiber mats containing silver nanoparticles. Polymer. 2008;49(21):4723–4732. DOI: 10.1016/j.polymer.2008.08.021.

[23] Hong KH. Preparation and properties of electrospun poly(vinyl alcohol)/silver fiber web as wound dressings. Polym Eng Sci. 2007;47(1):43–49. DOI: 10.1002/pen.20660.

[24] Park JH, Karim MR, Kim IK, Cheong IW, Kim JW, Bae DG, Cho JW, Yeum JH. Electrospinning fabrication and characterization of poly(vinyl alcohol)/montmorillon-

ite/silver hybrid nanofibers for antibacterial applications. Colloid Polym Sci. 2010;288:115–121. DOI: 10.1007/s00396-009-2147-4.

[25] An J, Zhang H, Zhang J, Zhao Y, Yuan X. Preparation and antibacterial activity of electrospun chitosan/poly(ethylene oxide) membranes containing silver nanoparticles. Colloid Polym Sci. 2009;287(12):1425–1434. DOI: 10.1007/s00396-009-2108-y.

[26] Penchev H, Paneva D, Manolova N, Rashkov I. Electrospun hybrid nanofibers based on chitosan or *N*-carboxyethylchitosan and silver nanoparticles. Macromol Biosci. 2009;9:884–894. DOI: 10.1002/mabi.03.

[27] Xu X, Yang Q, Wang Y, Yu H, Chen X, Jing X. Biodegradable electrospun poly(l-lactide) fibers containing antibacterial silver nanoparticles. Eur Polym J. 2006;42(9):2081–2087. DOI: 10.1016/j.eurpolymj.2006.03.032.

[28] Augustine R, Kalarikkal N, Thomas S. Electrospun PCL membranes incorporated with biosynthesized silver nanoparticles as antibacterial wound dressings. Appl Nanosci. 2016;6(3). DOI: 10.1007/s13204-015-0439-1.

[29] Park JH, Park SM, Kwon IJ, Hyun JM, Deng Y, Jeong YG, Cheong IW, Yeum JH. Poly(vinyl alcohol)/montmorillonite/silver hybrid nanoparticles prepared from aqueous solutions by the electrospraying method. J Compos Mater. 2013;47(27):3367–3378. DOI: 10.1177/0021998312465100.

[30] Oskam G. Metal oxide nanoparticles: synthesis, characterization and application. J Sol–Gel Sci Technol. 2006;37(3):161–164. DOI: 10.1007/s10971-005-6621-2.

[31] Kim I-D, Rothschild A. Nanostructured metal oxide gas sensors prepared by electrospinning. Polym Adv Technol. 2011;22(3):318–325. DOI: 10.1002/pat.1797.

[32] Seabra A, Durán N. Nanotoxicology of metal oxide nanoparticles. Metals. 2015;5(2): 934–975. DOI: 10.3390/met5020934.

[33] Lee K, Lee S. Multifunctionality of poly(vinyl alcohol) nanofiber webs containing titanium dioxide. J Appl Polym Sci. 2012;124(5):4038–4046. DOI: 10.1002/app.34929.

[34] Son B, Yeom B-Y, Song SH, Lee C-S, Hwang TS. Antibacterial electrospun chitosan/poly(vinyl alcohol) nanofibers containing silver nitrate and titanium dioxide. J Appl Polym Sci. 2009;111(6):2892–2899. DOI: 10.1002/app.29233.

[35] Korina E, Stoilova O, Manolova N, Rashkov I. Multifunctional hybrid materials from poly(3-hydroxybutyrate), TiO_2 nanoparticles, and chitosan oligomers by combining electrospinning/electrospraying and impregnation. Macromol Biosci. 2013;13(6):707–716. DOI: 10.1002/mabi.201200410.

[36] Korina E, Stoilova O, Manolova N, Rashkov I. Poly(3-hydroxybutyrate)-based hybrid materials with photocatalytic and magnetic properties prepared by electrospinning and electrospraying. J Mater Sci. 2014;49(5):2144–2153. DOI: 10.1007/s10853-013-7907-3.

[37] Xie Y, He Y, Irwin PL, Jin T, Shi X. Antibacterial activity and mechanism of action of zinc oxide nanoparticles against *Campylobacter jejuni*. Appl Environ Microbiol. 2011;77(7):2325–2331. DOI: 10.1128/AEM.02149-10.

[38] Jones N, Ray B, Ranjit KT, Manna AC. Antibacterial activity of ZnO nanoparticle suspensions on a broad spectrum of microorganisms. FEMS Microbiol Lett. 2008;279(1): 71–76. DOI: 10.1111/j.1574-.01012.x.

[39] Xu X, Chen D, Yi Z, Jiang M, Wang L, Zhou Z, Fan X, Wang Y, Hui D. Antimicrobial mechanism based on H_2O_2 generation at oxygen vacancies in ZnO crystals. Langmuir. 2013;29(18):5573–5580. DOI: 10.1021/la400378t.

[40] Kołodziejczak-Radzimska A, Jesionowski T. Zinc oxide—from synthesis to application: a review. Materials. 2014;7(4):2833–2881. DOI: 10.3390/ma7042833.

[41] Shalumon KT, Anulekha KH, Nair SV, Chennazhi KP, Jayakumar R. Sodium alginate/ poly(vinyl alcohol)/nano ZnO composite nanofibers for antibacterial wound dressings. Int J Biol Macromol. 2011;49(3):247–254. DOI: 10.1016/j.ijbiomac.2011.04.005.

[42] Augustine R, Dominic EA, Reju I, Kaimal B, Kalarikkal N, Thomas S. Electrospun polycaprolactone membranes incorporated with ZnO nanoparticles as skin substitutes with enhanced fibroblast proliferation and wound healing. RSC Adv. 2014;4(47): 24777–24785. DOI: 10.1039/c4ra02450h.

[43] Augustine R, Malik HN, Singhal DK, Mukherjee A, Malakar D, Kalarikkal N, Thomas S. Electrospun polycaprolactone/ZnO nanocomposite membranes as biomaterials with antibacterial and cell adhesion properties. J Polym Res. 2014;21(3):1–17. DOI: 10.1007/ s10965-013-0347-6.

[44] Augustine R, Kalarikkal N, Thomas S. Effect of ZnO nanoparticles on the in vitro degradation of electrospun polycaprolactone membranes in simulated body fluid. Int J Polym Mater Polym Biomater. 2016;65(1):28–37. DOI: 10.1080/00914037.2015.1055628.

[45] Virovska D, Paneva D, Manolova N, Rashkov I, Karashanova D. Electrospinning/ electrospraying vs. electrospinning: a comparative study on the design of poly(l-lactide)/zinc oxide non-woven textile. Appl Surf Sci. 2014;311:842–850. DOI: 10.1016/ j.apsusc.2014.05.192.

[46] Heriberto R-T. Nuevos Materiales Híbridos Basados en Poliésteres Biodegradables y Nanopartículas de ZnO: Propiedades Antibacterianas y Protección UV [dissertation]. Saltillo, México: Centro de Investigación en Química Aplicada; 2015. 221 p.

[47] Ungur G, Hrůza J. Influence of copper oxide on the formation of polyurethane nanofibers. Fibers Polymers. 2015;16(3):621–628. DOI: DOI 10.1007/s12221-015-0621-9.

[48] Zhang C, Li C, Bai J, Wang J, Li H. Synthesis, characterization, and antibacterial activity of Cu NPs embedded electrospun composite nanofibers. Colloid Polym Sci. 2015;293(9): 2525–2530. DOI: 10.1007/s00396-015-3640-6.

Chapter 6

Novel Natural Polymer/Medicinal Plant Extract Electrospun Nanofiber for Cosmeceutical Application

Jeong Hyun Yeum, Sung Min Park,
Seong Baek Yang, Yeasmin Sabina,
Young Hun Kim and Jae Cheon Shin

Additional information is available at the end of the chapter

http://dx.doi.org/10.5772/63412

Abstract

Zein produced from maize is a hydrophobic protein, which holds great potential for a number of industrial applications, for example, food packaging, pharmaceutical, cosmetic, and biomedical industry. Sorghum, known as important cereal crop worldwide, is a good source of various phytochemicals such as tannins, phenolic acids, anthocyanins, phytosterols, and policosanols, and these phytochemicals have great impact on human health. *Poria cocos*, a well-known traditional East-Asian medicinal plant, is found around the roots of pine trees in Korea and China. As a rapid and efficient process, electrospinning has drawn huge interest among scientists to produce nanostructured polymer materials with excellent properties. In this work, we studied the influence of co-solvent ratio and concentration of zein/medicinal plant extract on the morphologies of nanostructured zein/medicinal plant extract nanomaterials prepared by electrospinning technique from ethanol/water solution. The zein/medicinal extract nanofibers were characterized by field-emission scanning electron microscopy, transmission electron microscopy, thermogravimetric analysis, and differential scanning calorimetric technique. And we were to incorporate medicinal plant resources into the electro-spun zein nanofibers by electrospinning technique to investigate the effect of medicinal extract on the morphologies, antibacterial, antioxidant, and other properties. Zein/medicinal plant extract might have a practical use as a new preservative for cosmeceutical applications.

Keywords: natural polymer, medicinal plant extract, sorghum extract, *Poria cocos* extract, Electrospinning, Nanofiber, Cosmeceutical, zein

1. Introduction

Recently, investigations have been focused on incorporating medicinal plant extracts with natural polymer-based electro-spun nanofibers due to their unique characteristics and applications. Such materials can be prepared by incorporating medicinal plant extract into the nanofibers of natural polymer by electrospinning technique. Electrospinning has generated great interest among scientists due to its very simple, low-cost method to produce nanofiber which has exhibited outstanding properties such as high porosity and a high specific surface area [1–5]. Biomaterials are of huge interest to the scientists due to their inclusive potential applications particularly in tissue engineering and drug delivery [6–8]. Among the various types of biomaterials being developed, electro-spun ultrafine fibers of protein-based biomaterials are preferred for their cosmeceutical and medical applications. Having smaller pores and higher surface area than general fibers, electro-spun fibers have been successfully used in different fields such as tissue engineering scaffolds, biomedical, pharmaceutical, healthcare, biotechnology, and others [9–13]. Medicinal plants act as a rich source for antimicrobial agents and used as drugs and cosmetics due to their medicinal properties [14–18].

The principle of electrospinning is simple; however, it is necessary to control the process as several variables have an influence on the properties of the end product [9, 19–22]. This chapter is concerned with the fabrication of natural polymer/medicinal plant extract electro-spun nanofibers for cosmeceutical application which is one of the most important application areas. Furthermore, the effects of co-solvent ratio and concentration of natural polymer/medicinal plant extract on the morphologies of nanostructured natural polymer/medicinal plant extract electro-spun nanofibers were also investigated. The effectiveness of these bio-nanofibers is demonstrated with a field emission scanning electron microscope (FE-SEM), a transmission electron microscopy (TEM), thermogravimetric analysis (TGA), a differential scanning calorimetry (DSC) and the anti-bacterial performance and antioxidant properties were also discussed.

2. Effect of co-solvent ratios on the morphologies of electro-spun zein/sorghum extract and zein/*Poria cocos* extract nanomaterials

Zein is considered as one of the best understood plant proteins and it is soluble in ethanol/water solution. It presents good cell compatibility and has more hydrophobic characteristics than other proteins as a result of the presence of polar amino acids, proline and glutamine, which are the major constituents of zein [23, 24]. As a biocompatible and biodegradable protein, zein has useful applications in tissue engineering and drug delivery [8, 23–26]. Also, zein has been extensively used in food packaging, cosmetic, pharmaceutical and biomedical industries [24, 27–32]. Another useful plant protein sorghum (*Sorghum bicolor* L. Moench) is an important cereal food crop and in Asia and Africa, it is considered as an important human food resource [33]. It contains various phytochemicals, for example tannins, phenolic acids, anthocyanins, phytosterols, and policosanols and these phytochemicals have great potential to impact human health significantly [34]. According to recent studies, sorghum has antioxidant activity [35,

36], anti-carcinogenic effects [36, 37], cholesterol-lowering effects [38], antimicrobial effects [37] and it has a potential to reduce the risk of cardiovascular disease [35, 36]. *Poria cocos* is used in traditional East-Asian medicine that is found around the roots of pine trees in Korea and China. As a rapid and efficient process, electrospinning has attracted great interest among scientists to produce nanostructured polymer materials having excellent properties. Though the principle of electrospinning technique is simple; however, it is difficult to control the process and several variables influence the properties of end product [39].

There are many parameters which have great influence on the morphology of the resultant electro-spun materials varying from nanoparticles to nanofibers, having pores on their surface to beaded nanofibers. Electrospinning parameters can be broadly classified into polymer solution parameters and processing conditions that comprise the applied voltage, tip-to-collector distance (TCD), etc. In this work, we have studied the influence of co-solvent ratio and concentration of zein/sorghum extract and zein/*Poria cocos* extract on the morphologies of nanostructured zein/sorghum extract and zein/*Poria cocos* extract nanomaterials prepared by electrospinning technique from ethanol/water solution. The morphological changes were observed in details by FE-SEM and TEM.

2.1. Experimental

2.1.1. Materials

Zein (molecular weight = 35,000) extracted from corn was purchased from Tokyo Chemical Industry Co. Ltd., Japan. Sorghum was collected from National Institute of Crop Science (NICS) (Rural Development Administration, Korea). *Poria cocos* was obtained from cultivation in a plastic bag. Double-distilled water was used with 96% (v/v) ethanol (EtOH) (obtained from Daejung Chemical & Materials Co. Ltd., Korea) as a co-solvent to prepare all solutions.

2.1.2. Preparation of ethanolic extract and water extract of sorghum and Poria cocos

In case of ethanolic extract, at first, the stem and leaves of sorghum (*Sorghum bicolor* L. Moench) and the sliced *Poria cocos* were air dried and chopped. Later, they were extracted three times using 70% ethanol for 14 days at room temperature. The extracts were combined and evaporated for drying under reduced pressure, and temperature was kept bellow 40°C to yield a red gum. In case of water extract, the stem and leaves of sorghum and the air-dried *Poria cocos* were extracted with distilled water by refluxing for 24 hours (three times) on a water bath at 85°C. The extract was filtered through a Buchner funnel using filter paper, concentrated by rotary evaporator, and freeze dried.

2.1.3. Preparation of spinning solution

Zein solutions were obtained by dissolving zein at different volume ratios of ethanol/water [9:1, 8:2, and 7:3 (v/v)] and various zein concentrations were used (10, 15, 20, 25, and 30 wt.%). Zein/sorghum extract blend solutions were prepared by dissolving various amounts of sorghum extracts in 7:3 (v/v) mixture solvent of EtOH/H_2O at room temperature for 10 minutes.

After that, 30 wt.% zein (based on the weight of the solution) was added into the above solution and the mixture was stirred for another 10 minutes. Following this similar method, zein/*Poria cocos* solutions were also prepared for electrospinning.

2.1.4. Electrospinning

Zein solutions were carefully placed into a syringe and a syringe pump was used to deliver the solution through the blunt needle with a controlled solution feeding rate. Electrospinning was carried out under a constant electric field of 10 kV (Chungpa EMT Co., Korea), which was applied to the solution via an alligator clip attached to the syringe needle. An electrically grounded Al foil was used to collect fibers, which was placed at 15 cm vertical distance to the needle tip. Same method as described here was used for electrospinning zein/sorghum extract and zein/*Poria cocos* blend solutions. **Figure 1** shows a schematic diagram of the electrospinning process.

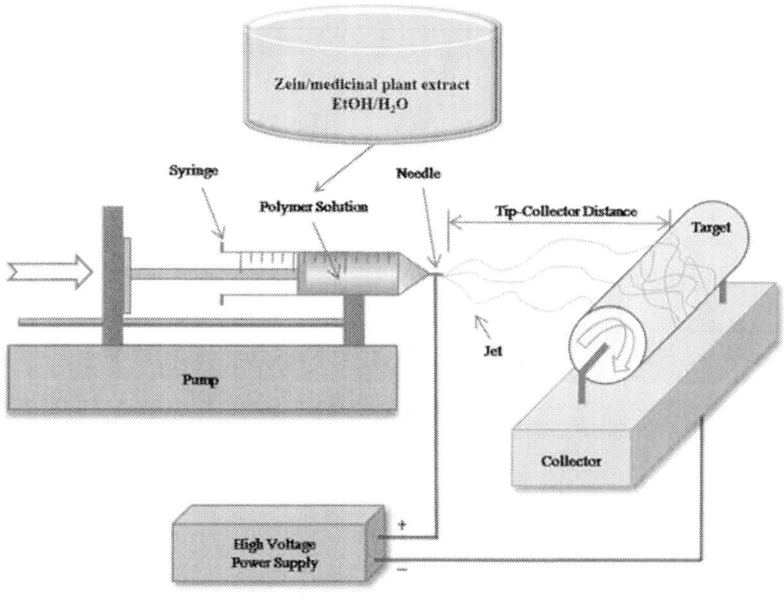

Figure 1. Schematic representation of the electrospinning process.

2.1.5. Characterization

FE-SEM images of the electro-spun nanofibers of zein/*Poria cocos* extract were captured using microscope (JSM-6380, JEOL Ltd., Japan) after platinum coating. The TEM (H-7600, Hitachi Ltd., Japan) analysis was conducted with an accelerating voltage of 100 kV.

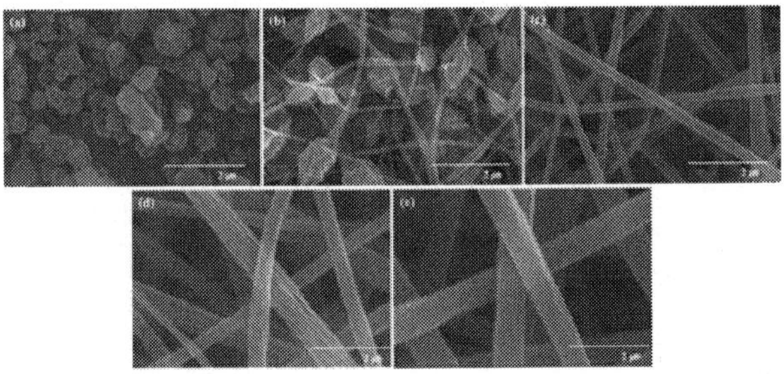

Figure 2. FE-SEM images of electro-spun zein nanofibers prepared by different zein concentrations of (a) 10 wt.%, (b) 15 wt.%, (c) 20 wt.%, (d) 25 wt.%, and (e) 30 wt.% [EtOH/H_2O ratio of 70:30 (v/v), TCD = 15 cm and applied voltage = 10 kV].

Figure 2 shows the variation in the FE-SEM morphologies of the 10, 15, 20, 25, and 30 wt.% of zein electro-spun nanofibers from the 7:3 (v/v) of ethanol/water co-solvents. At a fixed applied voltage (10 kV) and TCD (15 cm), the effects of the polymer solution concentration on the morphology of zein nanofiber are illustrated. At low concentration of polymer solution (10 wt. % and 15 wt.%), mostly beads were produced as shown in **Figures 2a** and **2b**. However, when the zein concentration was increased, fibrous products started getting formed and beads density decreased gradually (**Figures 2c** and **2d**). Finally, uniform zein nanofiber mats having no bead defects were observed at 30% zein concentration (**Figure 2e**). Ultrafine zein electro-spun nanofiber mats in a nanometer range (300–500 nm) prepared in the aqueous solutions are presented in **Figure 2e**. The key factors that control the formation of the beaded fibers are charge density carried by the jet, the viscoelasticity, and surface tension of the solution [21]. The effect of the different volume ratios of ethanol/water in the polymer solution was also evaluated by changing the ethanol content from 70% to 90% (v/v) and the results are presented in **Figures 3** and **4**. However, the most regular-diameter ultrafine nanofibers were found at EtOH/water ratio of 70:30 and 30 wt.% zein concentration (**Figure 2e**).

Figure 3. FE-SEM images of electro-spun zein nanofibers prepared by different zein concentrations of (a) 10 wt.%, (b) 15 wt.%, (c) 20 wt.%, (d) 25 wt.%, and (e) 30 wt.% [EtOH/H_2O ratio of 80:20 (v/v), TCD = 15 cm and applied voltage = 10 kV].

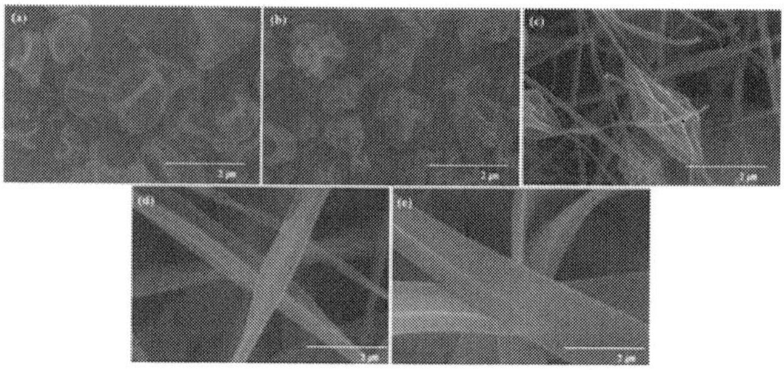

Figure 4. FE-SEM images of electro-spun zein nanofibers prepared by different zein concentrations of (a) 10 wt.%, (b) 15 wt.%, (c) 20 wt.%, (d) 25 wt.%, and (e) 30 wt.% [EtOH/H_2O ratio of 90:10 (v/v), TCD = 15 cm and applied voltage = 10 kV].

FE-SEM morphologies of electro-spun zein nanofibers containing various quantities of sorghum (0, 5, 10, and 20 wt.%) and *Poria cocos* extract (0, 5, 10, and 15 wt.%) are presented in **Figure 5** and **6**, respectively. At a fixed applied voltage (10 kV) and TCD (15 cm), the typical morphology of only zein nanofibers is shown in **Figure 5a**. The morphologies of the nanofibers were gradually changed when the amount of sorghum extract was increased from 0 to 20 wt. %. A slight reduction in the fiber diameters was observed with the increase of sorghum extract content. In **Figure 5b**, it can be clearly seen that nanofibers were splattered as 5 wt.% sorghum extract was added to zein solution. With the increase of the amount of sorghum extract, splashing of the nanofibers was increased and they started forming branched nanofibers that were connected with others (**Figures 5c** and **5d**).

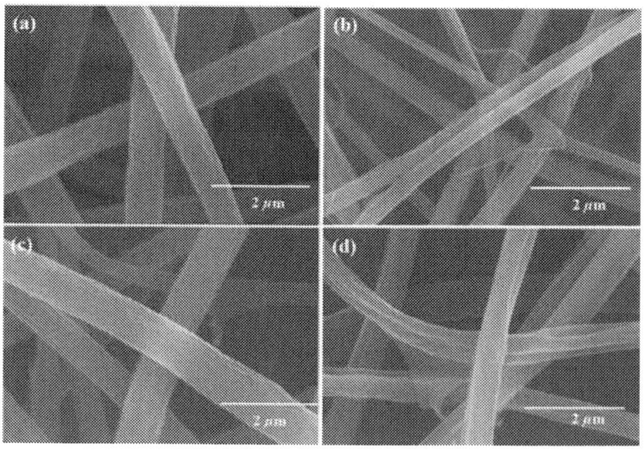

Figure 5. FE-SEM images of electro-spun zein nanofibers containing different sorghum extracts: (a) 0 wt.%, (b) 5 wt.%, (c) 10 wt.%, and (d) 20 wt.% [zein concentration = 30 wt.%, EtOH/H_2O of 7:3 (v/v), TCD = 15 cm and applied voltage = 10 kV].

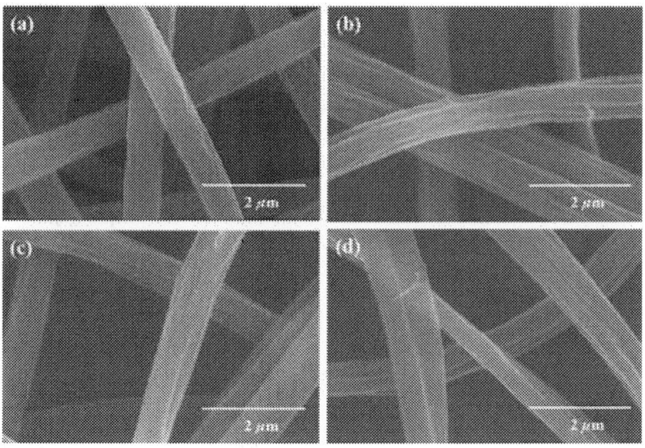

Figure 6. FE-SEM images of electro-spun zein nanofibers containing different *Poria cocos* extracts: (a) 0 wt.%, (b) 5 wt. %, (c) 10 wt.%, and (d) 15 wt.% [zein concentration = 30 wt.%, EtOH/H_2O ratio of 7:3 (v/v), TCD = 15 cm and applied voltage = 10 kV].

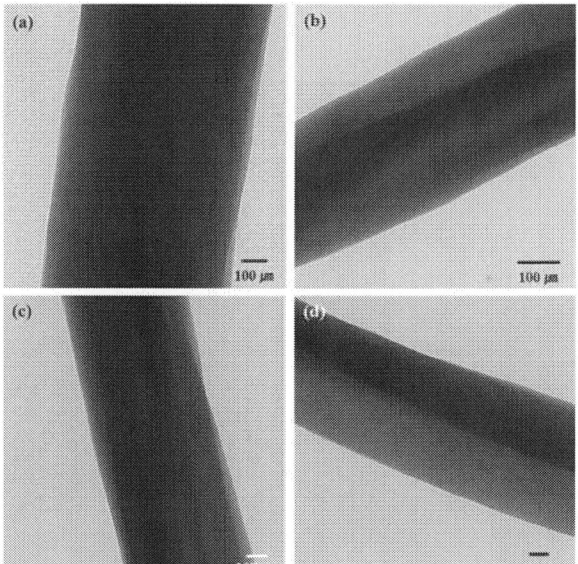

Figure 7. TEM images of electro-spun zein nanofibers containing different sorghum extracts: (a) 0 wt.%, (b) 5 wt.%, (c) 10 wt.%, and (d) 20 wt.% [zein concentration = 30 wt.%, EtOH/H_2O of 7:3 (v/v), TCD = 15 cm and applied voltage = 10 kV].

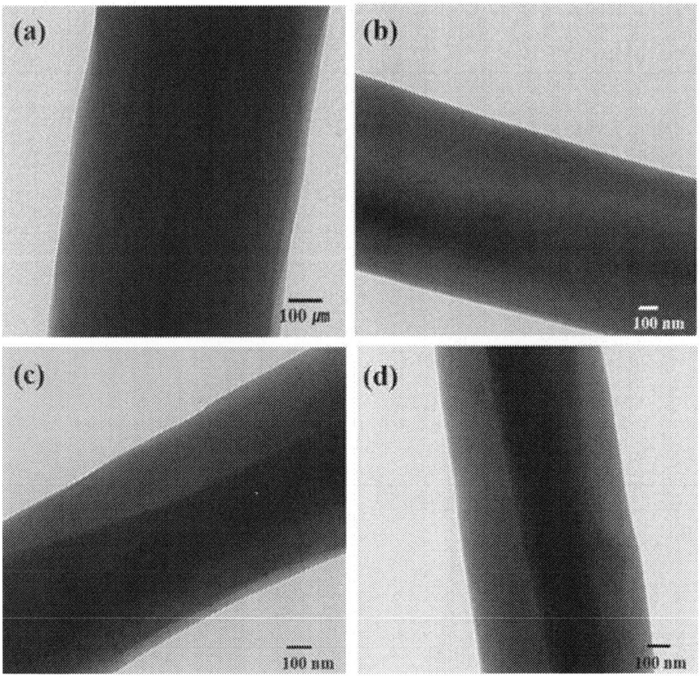

Figure 8. TEM images of electro-spun zein nanofibers containing different *Poria cocos* extracts: (a) 0 wt.%, (b) 5 wt.%, (c) 10 wt.%, and (d) 15 wt.% [zein concentration = 30 wt.%, EtOH/H_2O ratio of 7:3 (v/v), TCD = 15 cm and applied voltage = 10 kV].

The modified physical properties that developed in zein/sorghum extract solutions are responsible for these morphological changes [32]. An opposite trend was observed in case of zein/*Poria cocos* nanofibers as the fiber diameters expanded slightly with the increase of the amount of *Poria cocos* extract. Also, nanofibers showed the tendency of formation of rough surfaces when the *Poria cocos* extract was added; however, nanofibers containing 15 wt.% *Poria cocos* showed the maximum surface smoothness in comparison to those with 5 and 10 wt.% solutions (**Figures 6b**, **6c**, and **6d**). For preparing both types of electro-spun nanofibers (zein/sorghum extract and zein/ *Poria cocos* extract), a 7:3 (v/v) mixture solvent of EtOH/H_2O was used. The morphological changes of zein/sorghum extract and zein/ *Poria cocos* extract were also observed from their TEM images listed in **Figures 7** and **8**, respectively. The TEM images clearly show that both zein/sorghum extract and zein/ *Poria cocos* extract nanofibers were crinkled with the increase of sorghum extract and *Poria cocos* extract, respectively. However, folding is slight for *Poria cocos* extract in comparison to sorghum extract (**Figures 7b**, **7c**, **7d**, **8b**, **8c**, and **8d**). From FE-SEM and TEM images, it can be clearly observed that sorghum extract and *Poria cocos* extract could alter the zein morphology very effectively.

2.2. Effect of extraction conditions on yield

Extract yield of sorghum and *Poria cocos* using water and 70% EtOH is summarized in **Table 1**. For indicating the effects of the extraction conditions, extraction yield was used. In case of sorghum, 70% EtOH extract yield rate (1.414%) was maximum followed by water. For *Poria cocos*, the trend was similar as observed in the case of sorghum.

Medicinal plant extracts		Total (g)	Yield rate (%)
Water extract	Sorghum	0.9468	1.1482
	Poria cocos	0.8259	0.9177
70 % EtOH extract	Sorghum	1.5846	1.7414
	Poria cocos	1.3820	1.5356

Table 1. Yield rate of the extracts of sorghum and *Poria cocos* using water and ethanol.

3. Novel zein/medicinal plant extract electro-spun nanofibers for cosmeceutical application

Due to biocompatible nature, bio-nanofibers have superiority over their synthetic counterparts and have wide range of applications. Among the various types of bio-nanofibers being developed, extensive efforts are currently being made for the development of protein-based bio-nanofibers. Nanofibers have a large surface-to-mass ratio that makes them promising candidates for advanced material devices [40]. A number of methods have been developed to spin nanofibers, among them electrospinning is one of the methods using electrical charge to draw nanoscale fibers from polymer liquid solutions [20, 41]. Natural nanofibers are preferred products over synthetic nanofibers as synthetic nanofibers are environmentally toxic [40].

The major purpose of this section is the incorporation of medicinal plant resources (sorghum extract and *Poria cocos* extract) into the electro-spun zein nanofibers by electrospinning technique to investigate the effects of medicinal extract on the morphologies, antibacterial and antioxidant properties of zein nanofibers. The prepared zein/medicinal extract nanofibers were characterized by FE-SEM, TGA, and DSC techniques. Photoshop 5 was used to measure the fiber diameter. Using 2,2-diphenyl-2-picrylhydrazyl hydrate (DPPH) assay and superoxide dismutase (SOD) assay, antioxidant properties of Zein/*Poria cocos* plant extract were investigated and improved antioxidant capacities were found. Zein nanofibers containing sorghum extract were found to be effective to control the growth of bacteria. The significant anti-bacterial activity of zein/sorghum extract nanofibers and improved antioxidant capacities of zein/*Poria cocos* extract nanofibers suggested that they might have a practical use of new preservative for cosmeceutical and medical applications.

3.1. Experimental

3.1.1. Materials

Zein (molecular weight = 35,000) extracted from corn was purchased from Tokyo Chemical Industry Co. Ltd., Japan. Sorghum was collected from National Institute of Crop Science (NICS) (Rural Development Administration, Korea). *Poria cocos* was obtained from plastic bag cultivation method. Double-distilled water was used with 96% (v/v) EtOH (Daejung Chemical & Materials Co. Ltd, Korea) as a co-solvent to prepare electrospinning solution.

3.1.2. Electrospinning

Zein/medicinal plant resources (sorghum extract and *Poria cocos* extract) blend solution was carefully loaded into a syringe and the solution was delivered through blunt needle tip with a controlled flow rate by means of a single syringe pump. During electrospinning, the voltage was kept at 10 kV and applied to the solution via an alligator clip adhered to the syringe needle. Fibers were deposited on an electrically grounded Al foil and the distance between the needle and collector was maintained at 15 cm [42, 43].

3.1.3. Determination of antioxidant capacity

3.1.3.1. Free radical scavenging by the use of DPPH radical

The free radical scavenging activity was measured using DPPH radical following the protocols of Brand Williams modified by Miliauskas [44, 45]. Maximum absorption of DPPH radicals is at 515 nm and by an antioxidant compound, DPPH disappears with reduction. The DPPH solution in methanol (6×10^{-5} M) was required to prepare daily and 3 ml of this solution was mixed with 100 μl of methanolic solutions of *Poria cocos* extracts. In a water bath, the samples were incubated for 20 minutes at 37°C, and then the decrease in absorbance was measured at 515 nm (β). The measurement was carried out in triplicate and based on the following formula, DPPH activity was calculated.

$$\text{Inhibition}\,\% = [(\alpha - \beta)] \times 100$$

where, α= absorbance of the blank sample and β = absorbance of the *Poria cocos* extract.

3.1.3.2. Free radical scavenging by the superoxide dismutase (SOD) assay

The superoxide anion scavenging activity of *Poria cocos* extract was determined by the WST (2-(4-iodophenyl)-3-(4-nitrophenyl)-5-(2,4-disulphophenyl)2H-tetrazolium, monosodium salt) reduction method, using SOD assay kit-WST. In this method, O^2 reduced WST-1 and produced yellow formazan which is spectrophotometrically measured at 450 nm. Antioxidants are capable to inhibit yellow WST formation. All experiments were done in triplicate. The percentage of inhibition of superoxide radicals was measured using the formula discussed earlier.

3.1.4. *Determination of anti-microbial efficacy*

For investigating anti-bacterial performance of zein/medicinal plant extract, *Klebsiella pneumonia* (ATCC 4352) was used. To prepare samples, nanofibers were dispersed into a viscous aqueous solution containing 0.05 wt.% of polyoxyethylenesorbitan monooleate (TWEEN 80, Sigma Aldrich). After 18 hours of incubation at 37°C, a mixed culture of microorganisms was obtained on tryptone soya broth. After that, 0.4 g of sample was inoculated with 0.2 g of the microorganism suspension to adjust the initial concentration of bacteria to 1.3×10^5 ea/ml. Finally, the inoculants were mixed homogeneously with the samples and stored at 37°C.

3.1.5. *Characterization*

Average fiber diameters of the zein/*Poria cocos* extract and zein/sorghum extract nanofiber were determined using Photoshop 5. The thermal properties of electro-spun zein/sorghum and zein/*Poria cocos* extract were studied with TGA (model Q-50) and DSC (Q-10) from TA instruments under the nitrogen gas atmosphere. Using DPPH radical scavenging assay and SOD assay, antioxidant properties of zein/*Poria cocos* plant extract were investigated. The anti-bacterial performance test of zein/sorghum extract was carried out on *Klebsiella pneumonia*.

3.2. Results and discussion

3.2.1. *Morphology*

Figure 9 presents the dependence of the average diameters of the zein/sorghum and zein/*Poria cocos* extract nanofibers on the contents of sorghum and *Poria cocos* extracts, respectively. The result indicates a dramatic decrease in fiber diameter when 5 wt.% sorghum extract was added to zein nanofiber. However, nanofibers retained their previous shape (average diameter around 600 nm) when 10 wt.% sorghum extract was added. Moreover, there is no huge change in fiber diameter, when the sorghum extract contents increased from 10 to 20 wt.%. On the other hand, when the *Poria cocos* extract contents increased from 5 to 15 wt.%, a slight increase in fiber diameter was observed.

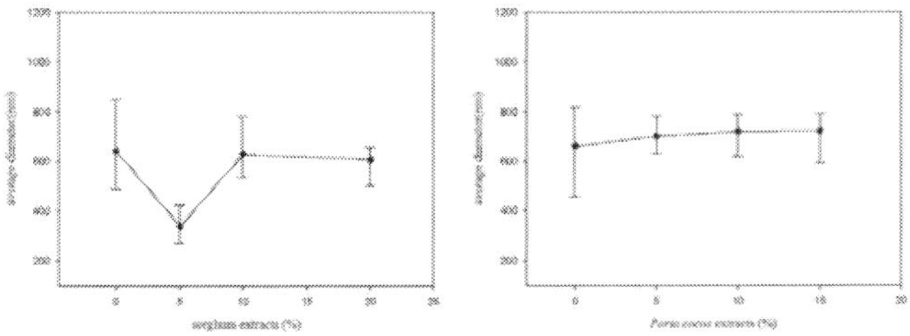

Figure 9. Average diameter of zein/sorghum extract and zein/*Poria cocos* extract nanofibers containing various sorghum extracts and *Poria cocos* extracts, respectively [zein concentration = 30 wt.%, $EtOH/H_2O$ of 7:3 (v/v), TCD = 15 cm and applied voltage = 10 kV].

3.2.2. *Thermal properties*

Temperature		50-150°C	150-350°C	350-450°C	Above 500°C
Concentration		Dehydration and decomposition of the solvent	Decomposition of the sorghum extract	Decomposition of thezein	Residue
Zein	Sorghum extract				
30 wt.%	-	92.12%	34.17%	20.25%	18.51%
30 wt.%	5 wt.%	92.94%	34.62%	21.81%	19.35%
30 wt.%	10 wt.%	93.49%	38.15%	26.48%	24.64%
30 wt.%	20 wt.%	94.87%	43.65%	30.03%	27.72%

Table 2. TGA data of electro-spun zein nanofibers containing different sorghum extracts: (a) 0 wt.%, (b) 5 wt.%, (c) 10 wt.%, and (d) 20 wt.% [zein concentration = 30 wt.%, $EtOH/H_2O$ of 7:3 (v/v), TCD = 15 cm and applied voltage = 10 kV].

To measure thermal properties of zein/sorghum extract and zein/*Poria cocos* extract nanofibers, TGA and DSC were used in nitrogen atmosphere. **Table 2** presents the TGA data of decomposition temperature obtained from zein nanofibers containing 0, 5, 10, and 20 wt.% of sorghum extract. From **Table 2**, it can be seen that, zein/sorghum extract nanofiber showed three typical weight loss regions. The first region (50–150°C) was due to dehydration and decomposition of the solvent, the second region (150–350°C) was due to the decomposition of the sorghum extract and the third region (350–450°C) was due to thermal degradation of zein. After thermal decomposition, the lowest amount of residue was obtained from pure zein nanofibers (18.51%) and highest amount of residue was obtained from zein nanofiber containing 20 wt.% of sorghum extract. It can be concluded that higher thermal properties could be obtained at higher weight percentage of sorghum extract. Similar effect was observed in case of zein/*Poria cocos* nanofibers as shown in **Table 3**.

Temperature		50-150°C	150-350°C	350-450°C	Above 500°C
Concentration		Dehydration and decomposition of the solvent	Decomposition of the *Poria cocos* extract	Decomposition of the zein	Residue
Zein	*Poria cocos* extract				
30 wt.%	-	92.24%	35.26%	19.16%	17.15%
30 wt.%	5 wt.%	92.81%	35.71%	20.84%	18.45%
30 wt.%	10 wt.%	93.24%	37.66%	23.14%	22.16%
30 wt.%	20 wt.%	94.11%	39.72%	29.18%	25.72%

Table 3. TGA data of electro-spun zein nanofibers containing with different *Poria cocos* extracts: (a) 0 wt.%, (b) 5 wt.%, (c) 10 wt.% and (d) 20 wt.% (zein concentration = 30 wt.%, $EtOH/H_2O$ of 7:3 (v/v), TCD = 15 cm and applied voltage = 10 kV).

Zein/sorghum extract concentration	T_g (°C)
(a) 30 wt.%	155.60
(b) 30 wt.%/5 wt.%	159.32
(c) 30 wt.%/10 wt.%	160.47
(d) 30 wt.%/20 wt.%	166.77

Table 4. T_g values of zein and zein/sorghum extract fibers containing different sorghum extracts: (a) 0 wt.%, (b) 5 wt.%, (c) 10 wt.% and (d) 20 wt.% [zein concentration = 30 wt.%, EtOH/H_2O of 7:3 (v/v), TCD = 15 cm and applied voltage = 10 kV].

The DSC data of electro-spun zein nanofibers containing different sorghum extracts are shown in **Table 4**. The lowest glass transition temperature (155.60°C) and melting temperature were obtained from only zein nanofiber. Both glass transition temperature and melting temperature of zein/sorghum extract nanofibers were shifted towards higher temperature values with the higher weight percentage of sorghum extract. Similar effect was also observed when the *Poria cocos* extract was added to zein nanofiber (**Table 5**). These results suggest that sorghum extract and *Poria cocos* extract have an influence on the thermal properties of zein nanofibers.

Zein/*Poria cocos* extract concentration	Tg (°C)
(a) 30 wt.%	155.60
(b) 30 wt.%/5 wt.%	158.12
(c) 30 wt.%/10 wt.%	159.48
(d) 30 wt.%/15 wt.%	161.37

Table 5. T_g values of zein and zein/*Poria cocos* extract fibers containing different *Poria cocos* extracts: (a) 0 wt.%, (b) 5 wt. %, (c) 10 wt.% and (d) 15 wt.% [zein concentration = 30 wt.%, EtOH/H2O of 7:3 (v/v), TCD = 15 cm and applied voltage = 10 kV].

3.2.3. Antioxidant activities

Radical scavenging capacities of *Poria cocos* extract were determined using DPPH and SOD assay and results are presented in **Tables 6** and **7**. 15 wt.% *Poria cocos* extract showed the highest antioxidant capacity (98.69 ± 1.49%) of DPPH inhibition, followed by 10 wt.% *Poria cocos* extract (97.75 ± 1.42%), and 5 wt.% *Poria cocos* extract (96.88 ± 0.91%) which is the lowest antioxidant capacity in this assay.

Sample	Concentration (%)	Scavenging activity (%)
Poria cocos	5	96.88 ± 0.91
	10	97.75 ± 1.42
	15	98.69 ± 1.49

Table 6. DPPH radical scavenging activities of ethanol extract from *Poria cocos*.

Sample	Concentration (%)	SOD activity (%)
Poria cocos	5	66.51 ± 6.45
	10	77.57 ± 4.34
	15	83.49 ± 5.73

Table 7. SOD activities of ethanol extract from *Poria cocos.*

Results obtained using SOD assay differed from those observed for DPPH assays. 15 wt.% *Poria cocos* showed the highest antioxidant capacities (83.49 ± 5.73%) followed by 10 wt.% *Poria cocos* extract (77.57 ± 4.34%). In this assay, 5 wt.% *Poria cocos* showed the lowest antioxidant potential (66.51 ± 6.45%). Therefore, both DPPH and SOD assay results suggest that antioxidant capacity increased with the increased weight percentage of *Poria cocos* extract.

3.2.4. *Anti-bacterial ability*

The anti-microbial performance of zein and zein/sorghum extract nanocomposite against *Klebsiella pneumonia* (ATCC 4352) was evaluated and results are shown in **Figure 11**. The anti-bacterial potency of only zein and zein/sorghum extract nanocomposite was assessed by counting the number of remaining bacteria with the storage time at 25°C. As shown in **Figure 11**, only zein nanocomposite has the anti-bacterial ability; however, in the test sample, a number of bacteria remained constant for a long time.

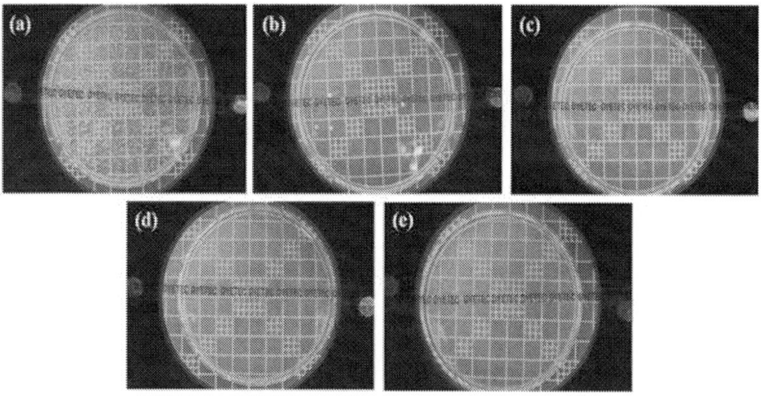

Figure 10. Antibacterial ability test with *Klebsiella pneumonia* (a) blank and electro-spun zein nanofibers containing different sorghum extracts: (b) 0 wt.%, (c) 5 wt.%, (d) 10 wt.% and (e) 20 wt.% (after 4 days).

On the other hand, zein/sorghum extract nanocomposites presented a remarkable decrease in the number of bacteria (**Figure 11c, 11d**, and **11e**). These results recommended that only a small proportion of sorghum extract can make zein more competent against bacteria. The increase in the concentration of the sorghum extract stimulates decrease in the number of bacteria.

http://dx.doi.org/10.5772/63412

4. Conclusion

Different volume ratios (v/v) of ethanol/water solutions and zein concentration can affect the morphology of the electro-spun zein nanofibers. Uniform zein nanofiber mats having average diameter around 600 nm could be obtained from the 7:3 (v/v) of ethanol with zein concentration of 30 wt.%. Different amounts of sorghum and *Poria cocos* extracts have been successfully incorporated into zein nanofibers by electrospinning technique from 7:3 (v/v) EtOH/H_2O solutions. Both sorghum extract and *Poria cocos* extract have shown remarkable effect on the structures and morphology of zein/sorghum extract zein/*Poria cocos* extract nanofibers and caused splattering, crinkling, and decreasing the diameter of the zein nanofibers due to decreased viscosity. Moreover, zein/sorghum extract and zein/*Poria cocos* extract nanofibers showed improved thermal property. Antioxidant properties of different wt. % of *Poria cocos* extracts were investigated using DPPH assay and SOD assay and found that 15 wt.% *Poria cocos* extract possessed highest capacities in both methods used, and hence could be a possible rich source of natural antioxidants. An improved performance was achieved from zein/sorghum extract nanofibers against bacteria compared to only zein nanofibers. The significant anti-bacterial activity of zein/sorghum extract nanofibers and improved antioxidant capacities of zein/*Poria cocos* extract nanofibers recommended that they might have a practical application as a new preservative for cosmeceutical and medical applications.

Acknowledgements

This research was supported by Technology Commercialization Support Program (113042-3), Ministry of Agriculture, Food, and Rural Affairs. Also, this work was partially supported by Kyungpook National University Research Fund, 2015.

Author details

Jeong Hyun Yeum[1*], Sung Min Park[2], Seong Baek Yang[1], Yeasmin Sabina[1], Young Hun Kim[3,4] and Jae Cheon Shin[5]

*Address all correspondence to: jhyeum@knu.ac.kr

1 Department of Bio-fibers and Materials Science, Kyungpook National University, Daegu, Korea

2 Korea Dyeing and Finishing Technology Institute, Daegu, Korea

3 Gennolab Co. Ltd., Gyeongsan, Korea

4 Y.S.M International Co. Ltd., Shanghai, China

5 Pohang Center for Evaluation of Biomaterials, Pohang Techno Park, Pohang, Korea

References

[1] Park JH, Kim IK, Choi JY, Karim MR, Cheong IW, Oh W, Yeum JH. Electrospinning fabrication of poly(vinyl alcohol)/waterborne polyurethane/silver composite nanofibre mats in aqueous solution for anti-bacterial exploits. Polymer & Polymer Composites. 2011;19(9):753-761.

[2] Ramakrishna S, Fujihara K, Teo WE, Yong T, Ma Z, Ramaseshan R. Electrospun nanofibers: solving global issues. Materials Today. 2006;9(3):40-50. DOI: 10.1016/S1369-7021(06)71389-X.

[3] Rezaei A, Nasirpour A, Fathi M. Application of cellulosic nanofibers in food science using electrospinning and its potential risk. Comprehensive Reviews in Food Science and Food Safety. 2015;14:269-284. DOI: 10.1111/1541-4337.12128.

[4] Subbiah T, Bhat GS, Tock RW, Parameswaran S, Ramkumar SS. Electrospinning of nanofibers. Journal of Applied Polymer Science. 2005;96(2):557-569. DOI: 10.1002/app.21481.

[5] Frenot A, Henriksson MW, Walkenstrom P. Electrospinning of cellulose-based nanofibers. Journal of Applied Polymer Science. 2007;103(3):1473-1482. DOI: 10.1002/app.24912.

[6] Chakraborty S, Liao IC, Adler A, Leong KW. Electrohydrodynamics: A facile technique to fabricate drug delivery systems. Advanced Drug Delivery Reviews. 2009;61(12):1043-1054. DOI: 10.1016/j.addr.2009.07.013

[7] Li WJ, Cooper JA, Mauck RL, Tuan RS. Fabrication and characterization of six electrospun poly(α-hydroxy ester)-based fibrous scaffolds for tissue engineering applications. Acta Biomaterialia. 2006;2(4):377-385. DOI: 10.1016/j.actbio.2006.02.005.

[8] Reddy N, Yang Y. Potential of plant proteins for medical applications. Trends in Biotechnology. 2011;29(10):490-498. DOI: 10.1016/j.tibtech.2011.05.003.

[9] Bhardwaj N, Kundu SC. Electrospinning: a fascinating fiber fabrication technique. Biotechnology Advances. 2010;28(3):325-347. DOI: 10.1016/j.biotechadv.2010.01.004.

[10] Seema A, Joachim HW, Andreas G. Use of electrospinning technique for biomedical applications. Polymer. 2008;49(26):5603-5621. DOI: 10.1016/j.polymer.2008.09.014.

[11] Sill TJ, Recum HA. Electrospinning: applications in drug delivery and tissue engineering. Biomaterials. 2008;29(13):1989-2006. DOI: 10.1016/j.biomaterials.2008.01.011.

[12] Liang D, Hsiao BS, Chu B. Functional electrospun nanofibrous scaffolds for biomedical applications. Advanced Drug Delivery Reviews. 2007;59(14):1392-1412. DOI: 10.1016/j.addr.2007.04.021.

[13] Barnes CP, Sell SA, Boland ED, Simpson DG, Bowlin GL. Nanofiber technology: designing the next generation of tissue engineering scaffolds. Advanced Drug Delivery Reviews. 2007;59(14):1413-1433. DOI: 10.1016/j.addr.2007.04.022.

[14] Mahesh B, Satish S. Antimicrobial activity of some important medicinal plant against plant and human pathogens. World Journal of Agricultural Sciences. 2008;4(S):839-843.

[15] Abbasi AM, Khan MA, Ahmad M, Zafar M, Jahan S, Sultanaa S. Ethnopharmacological application of medicinal plants to cure skin diseases and in folk cosmetics among the tribal communities of North-West Frontier Province. Journal of Ethnopharmacology. 2010;128(2):322-335. DOI: 10.1016/j.jep.2010.01.052.

[16] Antignac E, Re G, Clouzeau J, Toutain H. Safety of botanical ingredients in personal care products/cosmetics. Food and Chemical Toxicology. 2011;49(2):324-341. DOI: 10.1016/j.fct.2010.11.022.

[17] Bakkali F, Averbeck S, Averbeck D, Idaomar M. Biological effects of essential oils – a review. Food and Chemical Toxicology. 2008;46(2):446-475. DOI: 10.1016/j.fct.2007.09.106.

[18] Skocibusic M, Bezic N, Dunkic V. Phytochemical composition and antimicrobial activities of the essential oils from Satureja subspicata Vis. growing in Croatia. Food Chemistry. 2006;96(1):20-28. DOI: 10.1016/j.foodchem.2005.01.051.

[19] Pillay V, Dott C, Choonara YE, Tyagi C, Tomar L, Kumar P, Toit LC, Ndesendo VMK. A review of the effect of processing variables on the fabrication of electrospun nanofibers for drug delivery applications. Journal of Nanomaterials. 2013;2013(2013):1-22/789289. DOI: 10.1155/2013/789289.

[20] Huang ZM, Zhang YZ, Kotaki M, Ramakrishna S. A review on polymer nanofibers by electrospinning and their applications in nanocomposites. Composites Science and Technology. 2003;63(15):2223-2253. DOI: 10.1016/S0266-3538(03)00178-7.

[21] Fong H, Chun I, Reneker DH. Beaded nanofibers formed during electrospinning. Polymer. 1999;40(16):4585-4592. DOI: 10.1016/S0032-3861(99)00068-3.

[22] Deitzel JM, Kleinmeyer J, Harris D, Tan NC. The effect of processing variables on the morphology of electrospun nanofibers and textiles. Polymer. 2001;42(1):261-272. DOI: 10.1016/S0032-3861(00)00250-0.

[23] Reddy N, Yang Y. Novel green composites using zein as matrix and jute fibers as reinforcement. Biomass and Bioenergy. 2011;35(8):3496-3503. DOI: 10.1016/j.biombioe.2011.04.044.

[24] Shukla R, Cheryan M. Zein: the industrial protein from corn. Industrial Crops and Products. 2001;13(3):171-192. DOI: 10.1016/S0926-6690(00)00064-9.

[25] Lawton JW. Zein: a history of processing and use. Cereal Chemistry. 2002;79(1):1-18. DOI: 10.1094/CCHEM.2002.79.1.1.

[26] Rabbani MM, Kim YH, Yeum JH. Effect of co-solvent ratios and solution concentrations on morphologies of electrospun zein nanomaterials. Current Research on Agriculture and Life Sciences. 2013;31(1):11-17.

[27] Choi JH, Kim YH, Yeum JH. Effect of ethanol/water solvent ratios on the morphology of zein nanofiber mats and their wettability. Textile Coloration and Finishing. 2011;23(4):227-232. DOI: 10.5764/TCF.2011.23.4.227.

[28] Miyoshi T, Toyohara K, Minematsu H. Preparation of ultrafine fibrous zein membranes via electrospinning. Polymer International. 2005;54(8):1187-1190. DOI: 10.1002/pi.1829

[29] Torres S, Gimenez E, Lagaron JM. Characterization of the morphology and thermal properties of zein prolamine nanostructures obtained by electrospinning. Food Hydrocolloid. 2008;22(4):601-614. DOI: 10.1016/j.foodhyd.2007.02.005.

[30] Yao C, Li X, Song T. Electrospinning and crosslinking of zein nanofiber mats. Journal of Applied Polymer Science. 2007;103(1):380-385. DOI: 10.1002/app.24619.

[31] Yao C, Li X, Song T. Fabrication of zein/hyaluronic acid fibrous membranes by electrospinning. Journal of Biomaterials Science Polymer Edition. 2007;18(6):731-742. DOI: 10.1163/156856207781034070.

[32] Torres S, Ocio MJ, Lagaron JM. Novel antimicrobial ultrathin structures of zein/chitosan blends obtained by electrospinning. Carbohydrate Polymers. 2009;77(2):261-266. DOI: 10.1016/j.carbpol.2008.12.035.

[33] Ryu HS, Kim J, Kim HS. Enhancing effect of Sorghum bicolor L. Moench (sorghum, su-su) extracts on mouse spleen and macrophage cell activation. Korean Journal of Food and Nutrition. 2006;19(2):176-182.

[34] Awika JM, Rooney LW. Sorghum phytochemicals and their potential impact on human health. Phytochemistry. 2004;65(9):1199-1221. DOI: 10.1016/j.phytochem.2004.04.001.

[35] Choi YM, Jeong HS, Lee JS. Antioxidant activity of methanolic extracts from some grains consumed in Korea. Food Chemistry. 2006;103(1):130-138. DOI: 10.1016/j.foodchem.2006.08.004

[36] Kil HY, Seong ES, Ghimire BK, Chung IM, Kwon SS, Goh EJ, Heo K, Kim MJ, Lim JD, Lee DY, Yu CY. Antioxidant and antimicrobial activities of crude sorghum extract. Food Chemistry. 2009;115(4):1234-1239. DOI: 10.1016/j.foodchem.2009.01.032.

[37] Kwak CS, Lim SJ, Kim SA, Park SC, Lee MS. Antioxidative and antimutagenic effects of Korean buckwheat, sorghum, millet and job's tears. Journal of Korean Society Food Science Nutrition. 2004;33(6):921-929. DOI: 10.3746/jkfn.2004.33.6.921.

[38] Ha TY, Cho IJ, Lee SH. Screening of HMG-CoA reductase inhibitory activity of ethanol and methanol extracts from cereals and legumes. Korean Journal of Food Science and Technology. 1998;30(1):224-229.

[39] Heikkila P, Derlund LS, Ki JU, Kettunen L, Harlin A. Exploitation of electric field in controlling of nanofiber spinning process. Polymer Engineering and Science. 2007;47(12):2065-2074. DOI: 10.1002/pen.20923.

[40] Ifuku S, Shervani Z, Saimoto H. Preparation of chitin nanofibers and their composites. In: Dufresne A, Thomas S, Pothen LA, editors. Biopolymer Nanocomposites. 1st ed. New Jersey: John Wiley & Sons, Inc. Hoboken, N. J; 2013. p. 11-31. DOI: 10.1002/9781118609958.ch2.

[41] Frenot A, Chronakis IS. Polymer nanofibers assembled by electrospinning. Current Opinion in Colloid & Interface Science. 2003;8(1):64-75. DOI: 10.1016/S1359-0294(03)00004-9.

[42] Karim MR, Lee HW, Kim R, Ji BC, Cho JW, Son TW, Oh WT, Yeum JH. Preparation and characterization of electrospun pullulan/montmorillonite nanofiber mats in aqueous solution. Carbohydrate Polymers. 2009;78(2):336-342. DOI: 10.1016/j.carbpol.2009.04.024.

[43] Lee HW, Karim MR, Ji HM, Choi JH, Ghim HD, Park SM, Oh W, Yeum JH. Electrospinning fabrication and characterization of poly(vinyl alcohol)/montmorillonite nanofiber mats. Journal of Applied Polymer Science. 2009;113(3):1860-1867. DOI: 10.1002/app.30165

[44] Brand-Williams W, Cuvelier ME, Berset C. Use of a free radical method to evaluate antioxidant activity. LWT-Food Science and Technology. 1995;28(1):25-30. DOI: 10.1016/S0023-6438(95)80008-5.

[45] Miliauskas G, Venskutonis PR, Beek TA. Screening of radical scavenging activity of some medicinal and aromatic plant extracts. Food Chemistry. 2004;85(2):231-237. DOI: 10.1016/j.foodchem.2003.05.007.

Chapter 7

Nanofibers in Cosmetics

Fulya Yilmaz, Gizem Celep and Gamze Tetik

Additional information is available at the end of the chapter

http://dx.doi.org/10.5772/64172

Abstract

Nanofibers form the broadest class of nanomaterials due to their unique properties. High surface-area-to-volume ratios, low diameters, high strength values, low basis weights, high porosities, and small pore sizes make them good candidates for numerous applications. Cosmetics are one of these important application areas. Excellent interaction with environment (e.g., skin or damaged tissue), increased loading capacity for agents, high liquid absorption capacities, high oxygen, and water vapor permeability values are provided by characteristic properties of nanofibers. They are used as therapeutics, facial masks, skin care, and renewal products. This chapter will provide an overview of nanofibers in cosmetics. A brief history of cosmetics, different types of nanomaterials used in cosmetics, nanofiber properties, and production methods are described in this chapter. Novel applications of nanofibers in cosmetics are also mentioned.

Keywords: nanofibers, cosmetics, facial mask, skin care

1. Introduction

To be admired and look beautiful are two basic phenomena that exist in human nature. Seeming beautiful and clear and feeling admiration increase people's self-confidence and make them satisfied in terms of psychology. Besides seeming good and well-groomed, requirement for personal care increases the popularity of usage of cosmetics. The use of cosmetics is based on the ancient age. Archaeological excavations have proved that cosmetics have been commonly used in ancient Egypt. Dyes and paints that are used for coloring the skin and oils and perfumes that are used for personal hygiene were the earliest cosmetic products [1]. Although cosmetics were used in ancient age, they have become widespread in the 1900s and during the World War II and more conscious approaches for cosmetics usage have begun to be shown [2,

3]. As in every field, with the developments in cosmetics, qualities of cosmetic products used nowadays are enhanced and diversities of these products are increased. Especially, insertion of nanomaterials (nanofibers, nanoparticles, nanoliposomes, nanopigments, etc.) into the structures of cosmetics has given a new point of view to the cosmetic sector [4, 5].

Ancient Greek word "*nanos*" that means dwarf is used to describe the physical magnitudes in orders of 10^{-9}. Nanomaterials can be defined as materials having controllable sizes of 100 nm or less. Nanomaterials with enhanced properties are obtained by the reason of the fact that sizes at nano scales affect the structural, mechanical, thermal, thermo-dynamical, kinetical, and electrical properties of materials [6]. Since the discovery of these nanomaterials, rapidly increasing interest has been shown by the academic world and this has led to the production of various nanomaterials for diverse sets of applications. One of these application areas is cosmetics.

The usage of nanomaterials in cosmetics has continued for 40 years and has become popular since 1990s. The first usage of nanomaterials in cosmetics has been seen in moisturizers with liposomes. Nowadays, nanomaterials are widely used in sunscreen creams, anti-aging creams, hair products (such as anti-hairloss, anti-hairgraying, etc.), facial masks, and toothpastes [4, 5]. Also, personal care cosmetic products such as soap, shampoo, bronzer, moisturizer, foundation, and lip sticks are other examples of cosmetics that uses nanostructured materials [7, 8]. Besides personal care, researchers have been focusing on skin health care during the last decades. Skin wound healing, drug/agent delivery to skin for anti-wrinkle, and anti-aging purposes, and artificial skin applications are some challenging nanomaterial applications in cosmetics.

2. Nanomaterials in cosmetics

An engineered nanomaterial is defined as a material that is intentionally produced as having at least one dimension between the range of 1 and 100 nm [9]. Nanomaterials' molecular structure and advanced interactions with the environment are the two important factors that affect preference of nanomaterials in cosmetics [10]. The nanomaterials used in cosmetics can be classified as nanoemulsions, nanoliposomes, nanopigments, and nanofibers. These are all used in sunscreens, skin creams, hair products, and oral hygiene products [11].

2.1. Nanoemulsions

Nanoemulsions are defined as transparent or translucent liquid dispersion systems which are thermodynamically or kinetically stable. They contain nanometric sized droplets of water and oil. A surfactant is also found in the dispersion. There are three types of nanoemulsions depending on the composition: oil in water nanoemulsion (continuous aqueous phase where oil droplets are dispersed in), water in oil nanoemulsion (continuous oil phase where water droplets are dispersed in), and bicontinuous nanoemulsion (system where oil and water are interdispersed in) [12–16]. High pressure homogenization, microfluidization, phase inversion, sonication, ultrasonic system, and jet disperser methods are some methods that are used for the preparations of nanoemulsions [14, 15]. Physicochemical characterization of nanoemulsions can be performed by particle size analysis, rheological measurements, refractive index, and surface tension tests [14].

There are some advantages of nanoemulsions than the other types of emulsions, such as macro- and microemulsions. The amount of surfactant is lower in nanoemulsions when compared to microemulsions [17]. The advantages in terms of optical, tactile, and texture properties make nanoemulsions attractive in cosmetics. The problems like inherent creaming, flocculation, coalescence, and sedimentation that occur in macroemulsions and even in microemulsions have not been met in nanoemulsions due to their small droplet sizes which enhance transparent view and fluid properties [15, 18–20]. Reducing the sizes of emulsions to nanometers provides an increase in the content of nourishing oil, and by this means, the transparency and the lightness of the formula are preserved. Also, allowing fragilely transport of the active ingredient (like vitamins) to the skin by means of higher surface area is another advantage [11]. Properties such as freshness, purity, simplicity, and rapid penetration make nanoemulsions valuable for the cosmetic industry [12, 18, 21–23].

Sun care products, moisturizing and anti-aging creams, conditioners, and lotions are some of the examples of nanoemulsions. Another attractive example of oil-in-water form of nanoemulsions is polyethylene glycol emulsifier-free emulsions that have low-viscosity. With the advent of its moisturizing effect and low viscosity oil in water emulsions, it is used in emulsion-based wet wipes for baby-care and make-up removal. In addition to these products, nanoemulsions have potentials for personal hair products. For dry hairs, extended effect provided by cationic nanoemulsions usage makes hair gain less brittle, nongreasy, and shiny [12, 15, 24, 25]. Also, coenzyme encapsulated nanoemulsions enhance the collagen synthesis of fibroblasts. Cell culture studies revealed that there was an increase in the secretion of collagen when studied with coenzyme added nanoemulsion [26].

2.2. Nanoliposomes

Nanoliposome, as a term, means nanoscaled bilayered lipid vesicles or liposomes in the low nanometer size range. Phospholipid molecules of components combine in an aqueous solution to give a colloidal structure. This structure preserves its nanometric size during storage and application period in virtue of polar and nonpolar regions of bilayer forming molecules.

Nanoliposomes have the same physical, structural, and thermodynamical properties and also have the same way of formation with liposomes. Formation of both depends on the hydrophilic–hydrophobic interaction between phospholipids and water molecules [27–29]. There are various methods to produce nanoliposomes. High-pressure homogenization is one of the techniques and is used to produce liposomes ranging from 20 to 50 nm in size. Some of the other methods for producing nanoliposomes are sonication technique, extrusion method, microfluidization, heating method, and Mozafari method. Characterization of nanoliposomes may be performed by electron microscopy, radiotracers, fluorescence quenching, ultrasonic absorption, electron spin resonance spectroscopy, and nuclear magnetic resonance spectroscopy. Visual appearance, size distribution, stability, zeta potential, lamellarity, and entrapment efficiency may be investigated [30, 31].

Encapsulation and delivery of bioactive agents can be achieved by nanoliposomes. These can be used in cosmetics, nutraceuticals, and pharmaceuticals. In cosmetic sector, nanoliposomes are important because of their nanosized structure. They do not clog skin pores and allow penetration of air and water soluble materials, while the many types of cosmetic products cause accumulation of oil layer on the skin and block passing of air, water, vitamin, etc. through the skin. Benefiting from these features, some trademarks claim that they produce cosmetics with immediate lifting effect. As a result, nanoliposomes give wrinkleless effect to the skin appearance [31–33].

2.3. Nanopigments

Dispersions of organic and inorganic nanopigments have been used in various applications such as printing, paints, coatings, cosmetics, and color filter arrays for the display industry. Nanopigments or pigment nanoparticles which differ from the bulk materials are functional nanomaterials that are used in the cosmetics industry with a gradually broadening trend. The nanoinorganic pigments having particle sizes less than 100 nm are considered as an insoluble and chemically and physically inert into the substrate or binders [34–36].

Nanopigments have the advantage of consisting ultrasmall particles, and certainly, the dispersion of ultrasmall particles in solvents is easier than the larger ones. Also, their small shape enables more stable dispersion because the gravity can be ignored with regard to the other forces like van der Waals forces. Another advantage is the ability of forming closed packed films. Capillary force of small particles is greater than that of large particles [34].

Nanopigments already exist in our natural environment. Titanium dioxide (TiO_2) and zinc oxide (ZnO) are the well-known examples of nanopigments. Since they have unique capacity to reflect UV light, they are commonly used in sunscreens. They protect human skin against damages like skin cancer arising from UV radiation [37]. Other examples in the cosmetic market are firming lotions, bronzers, exfoliant scrubs, eye liners, hair coloring products, and styling gels [38–40]. Nanoaluminum oxide having the feature of diffusing the light with concealers and mineral foundations provide smooth effect to wrinkled areas on the face. Mica-based pigments are also used in cosmetics since they give pearlescent effect [35, 36, 39].

2.4. Nanofibers

Nanofibers are the last and the largest of nanomaterials used in cosmetics and are discussed in detail in Section 3.

3. Nanofibers: definition and properties

Fiber is defined as a unit of matter characterized by length, fineness, and high ratio of length to thickness by the Textile Institute [41]. Nanofibers are characterized as nanomaterials that have at least one dimension, that is, 100 nm or less, according to the definition of American National Science Foundation [42, 43]. Nanofibers are classified as one-dimensional (1D)

http://dx.doi.org/10.5772/64172

nanomaterials (having a degree of freedom in one direction). Apart from the definition in textiles, nanofibers are defined as fibers that have a minimum ratio of length to thickness of 1000:1 [44].

Properties of nanofibers can be specified as having low unit length (diameter), high surface-area-to-volume ratio, high strength value, low basis weight, high porosity, and small pore size [45–48]. These make them indispensable in numerous applications.

Increasing the number of fibers in a unit volume can be achieved by decreasing the diameters of fibers and by migrating atoms from the bulk to the surface. This leads to an increase in surface-area-to-volume ratio, and hence the enhanced liquid absorption capacities and increased retention levels of functional groups, ions, particulates, or agents are obtained. High contact surface area between nanofibers and skin provides delivery of cosmetic agents to the deeper skin parts. All these mean advancement in the activities of materials including nanofibers instead of micro ones (**Figure 1**).

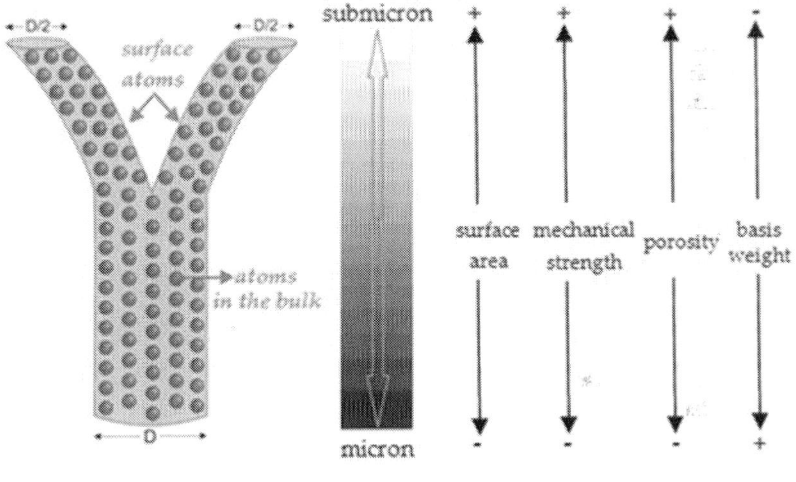

Figure 1. An exaggerated drawing of migration of atoms from bulk to surface and variation of properties with size.

The other properties enhanced by size reduction are the oxygen and water vapor permeability capabilities. According to high porosities and low pore diameters observed in nanofiber mats, breathable cosmetics can be successfully produced. This gains importance in terms of dermocosmetics that comprises cleaners, moisturers, and photoprotectors [49].

Mechanical strength of nanofibers is significant in terms of usage as artificial skin mats [50]. Since the mechanical properties of polymers are determined by molecular structure, weight, and orientation, polymer nanofibers that can be produced as highly molecularly oriented by fiber production methods serve as good templates in artificial skin applications. Low basis weight of nanofiber mats is the other great property.

4. Production methods of nanofibers

Nanofibers have been designed by choosing suitable polymers, convenient additives, and proper production methods to meet the requirements of their specific application area. Nanofibers are produced by lots of methods such as self-assembly, drawing, meltblowing, template synthesis, phase separation, melt spinning, centrifugal spinning, and electrospinning [51]. Flash spinning is a modified spunbond method and is also considered as one of the nanofiber production methods in some references [52]. The method is based on the formation of fibrillar formed filaments by extrusion through a spinneret and removal of solvent. But there is a lowest limit of about 1 μm for nanofiber diameters [53].

Although nanofibers can be achieved by the abovementioned methods, electrospinning is the unique method for the production of nanofibers in cosmetics. All the methods are described briefly except electrospinning, which is discussed elaborately because of its prevalence.

All the methods require sensitiveness at different levels due to the complexity of the process. Since the production is made in nanometer ranges, it should be avoided from atmospheric contaminations. If a visible threshold of about 40 μm is considered, the importance of clean process environment will be well understood.

4.1. Self-assembly

Self-assembly is the only nanofiber production method that benefit from bottom-up approach [54]. The term "self" corresponds to formation of molecules and molecular chains by atoms, finally forming fibers by molecular chains with minimum external effect [55]. The mechanism is based on formation of up structures (nanoscaled fibers) from bottom structures (molecules). The drawbacks of the method are complication of process and uncontrollable fiber diameters [56]. Leibmann et al. have found that self-assembled spider silk nanofibers and microbeads have potentials to be used in cosmetics and pharma [57].

4.2. Drawing

Drawing of individual nanofibers is achieved by a micromanipulator probe taking a little amount of polymer from the viscoelastic polymer solution droplet placed on a flat surface. Probe takes this little amount of polymer on and then draws it at a slow constant rate. By the contact of probe onto the flat surface at a distance adequate for fiber length, individual fibers are succeeded. Drawing rate and solvent evaporation rate are the important process parameters [56, 58, 59]. Polycaprolactone, polyethylene oxide, hyaluronic acid, fish gelatin blend, and polymethylmethacrylate are the biopolymer nanofibers formed by the drawing method [60], and in addition these, polymer nanofibers can be used as skin tissue scaffolds or skin wound healers.

4.3. Meltblowing

High flow rate of hot air is contacted with polymer extruding out of a capillary tip. Nanofibers are formed at a very short time [61]. The viscosity of the polymer melt should allow sufficient thinning of nanofibers in this process [62]. In this method, nanofibers of approx. 250 nm diameter can be obtained with low cost by using thermoplastic polymers [63]. The most common applications of meltblown nanofiber mats are filtering applications [64, 65]. Wet wipes constitutes the other class of applications [66, 67]. Also, a pack of cosmetic facial mask including a meltblown fiber mat layer was presented by Choi and Lee [68].

4.4. Template synthesis

As it is understood from the name of the method, templates are used for the production of nanofibers. Electrochemical, chemical, sol-gel, and chemical vapor deposition are some types of template synthesis [69]. Fiber diameters are determined by pore sizes of the templates. Since the process needs preliminary preparation (e.g., preparation of metal oxide nanoporous membranes), it is time consuming. Tao and Desai have pointed out the usage of biodegradable polymers for the production of template synthesized nanofibers for tissue engineering applications [70].

4.5. Phase separation

Basic principle of the phase separation is the formation of polymer-rich and polymer-poor phases and formation of nanofibrous structure after the removal of polymer-poor phase. A 3D fibrous structure is obtained instead of mats composed of individual fibers [56, 71].

Type of solvent and polymer, concentration of polymer solution, and temperature are the key parameters that affect the process [71]. Although there are limitations about the polymers that can be used in this method, there is an advantage of adjusting the mechanical properties by changing polymer concentration. Minimum equipment requirement is the other advantage [56, 72].

4.6. Melt spinning

In this continuous process, melt is drawn first and then fiber is wound with a faster rate than in extrusion [73]. Bicomponent spinning is a modified melt spinning method in which unconventional spinnerets are used. As it is obvious from the name of the method, two components are extruded through the spinneret, and then one component is removed and the other one remains [74]. Different spinneret designs may be used in the production. Islands in the sea model is the most used one [75]. Fibers approximately with diameters of 300 nm may be obtained by drawing 1 denier (g/9000 m) fibers from the island in the sea type spinneret [52].

4.7. Centrifugal spinning

Centrifugal spinning method is proposed as a method that exceeds the limitations of other nanofiber production methods such as available materials, production rate, safety, and cost [76]. A polymer melt is fed toward a spinning head (with a rotational speed of approx. 3000 rpm) which forms the centrifugal force that is required for sufficient attenuation [77]. There is an assistance of high velocity air during the transition of nanosize [76].Centrifugally spun biopolymer nanofibers were proposed as successful skin grafts with increased cell attachment and proliferation [78].

4.8. Electrospinning

Electrospinning is a simple peerless method that individual nanofibers can be formed continuously. It allows production of functionality added nanofibers from polymer solutions/melts to be used in various applications. Since electrospinning equipment is cheap and the process is simple, it is attractive for lots of researchers to produce nanofiber webs from a wide range of polymers.

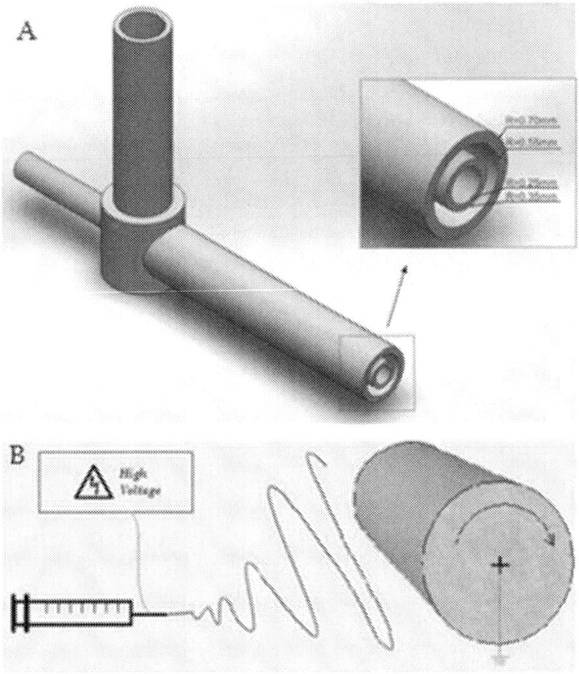

Figure 2. Schematic figure of (a) a coaxial needle and (b) an equipment with a rotating drum collector.

Basis of static electricity were laid by ancient Greeks. Thales of Miletus (630–550 BC) was the one who realized the force created by amber (electron) on objects. The first written evidence has been left by Theophrastus (374–287 BC) [79, 80]. William Gilbert has used an amber rod for drawing water droplets [80, 81]. From the 1700s lots of researchers investigated the behavior of droplets under electrostatic forces, electrospraying, stability of drawn jet, cone formation, and so on [51]. Producing silk-like synthetic fibers was an important topic in the 1900s. In 1934, Formhals took 11 patents about the subject. His mischance was the invention of polyamide

http://dx.doi.org/10.5772/64172

with a higher output than proposed in Formhals' methods [47]. As the academic interest increased by the time, the method was developed by modifications in feeding and collecting units as seen in **Figure 2a** and **b**, respectively. Coaxial needle provides feeding of two different polymers or encapsulation of agents, drugs, or other bioactive materials inside the shell polymer acting as a carrier. Rotating drum collectors develop the fiber structure, provide molecular orientation, and make electrospinning easy with high angular velocity.

Figure 3. A photo of an electrospun mat.

The principle of electrospinning is drawing of nanofibers under the electrostatic forces created with the usage of high voltage power supply and grounding placed on needle and collector, respectively [82–85]. The steps encountered in the process may be classified into six groups: the first one is the charging of polymer solution/melt causing a deformation on the droplet at the tip and the formation of cone [86, 87]. The second one is the generation of jet by increasing charging. This causes a deformation in jet [86–88]. At the third stage, the straight jet segment elongates [89, 90]. Deformation of straight jet segment, creation of instability, and continuation of elongation compose the fourth step [91, 92]. In the fifth step, solidification of nanofibers takes place by evaporation of solvent or cooling [87, 93]. And finally, fibers showing buckling behaviors are collected on the collector [94].

As one can understand from the six basic steps of electrospinning, the rate of feeding, electric field force acting on polymer solution/melt, and the flight distance of polymer jet are the key parameters of electrospinning process. It is known that none of the parameters can be evaluated independently from each other. In other words, there are optimum electrospinning parameters for every polymer solvent system or polymer melt. Since the adjustments of these parameters are easy, it can be said that the method still maintains its position at the focus of academic interest.

Producing nanofibers quite easily from various polymers makes electrospinning and electrospun products indispensible for a diverse set of applications. The electrospun nanofiber mats, as shown in **Figure 3**, may be used in drug delivery systems, therapeutics, and body care supplements [95].

5. Applications of nanofibers in cosmetics

Nanofibers take the advantage of their unique properties and have extensive usage in cosmetics, tissue engineering, biomedical, filtering, composites, protective clothing, electrical and optical applications, sensors, and agriculture [80, 96]. Since nanotechnology allows production of value-added products, cosmetics produced by nanotechnological methods have attracted attention from every area. With the aid of nanofiber production methods, especially by electrospinning, mats with controllable pore sizes and fiber diameters can be obtained. Also, the novel approaches that have been shown to cosmetics led to consumption of more conscious cosmetic products including therapeutic products and products for skin health and renewal (such as facial masks for skin cleansing, skin healing, and skin therapy) [95, 97]. All of the products mentioned above can be produced by using nanofibers. Applications of nanofibers in cosmetics are described below.

5.1. Facial masks and skin cleansings

Fathi-Azarbayjani et al. have presented an antioxidant and anti-wrinkle nanofiber face mask. They added ascorbic acid, retionic acid, collagen, and gold nanoparticles to the electrospinning solution and directly spun the fibers. In order to avoid from instability problems of prewetted facial masks, they have designed a mask that will be wetted just before the application. They indicated that wetting of masks will provide the release of agents from nanofibrous mats and ensure high skin penetration due to the high surface area of nanofibers. Thus, they developed healthy appearance of skin [98]. Smith et al. have a patent related to skin care mask. This mask is composed of nanofibers and can be used as a cleansing product for overaccumulated oil on skin [99]. Also, skin revitalizing factors can also be impregnated to nanofiber-based masks [95]. Kim designed a water soluble nanofiber layer for removal of cosmetic ingredients. He argued that dissolution of nanofiber layer by water maximized adhesion to the skin [100].

5.2. Skin health and renewal products

Researcher groups have attempted to electrospin nanofibers with various agents and investigate the effects on skin health. One of them produced cellulose acetate nanofibers by loading vitamin A and E to the spinning solution. They have searched the dermal therapeutic effect of vitamin loaded nanofiber mats as carriers. They compared the release of vitamins from the nanofiber mats and cast films and have not come across with a burst release in nanofibers as seen in cast films [101].

Another study including the loading of vitamin E to silk fibroin nanofibers was conducted in 2013. The researchers have argued that the mats presented in the study can be used as personal skin care products since they enhance the survival of the L929 fibroblast cells during *in vitro* tests [102].

5.3. Skin wound dressings and drug delivery products

Delivery of drugs to the skin can be considered in cosmetic applications, especially in dermocosmetics. Sufficient delivery results were observed due to the higher surface area of nanofibers [103]. A researcher group has prepared electrospun chitosan-based nanofiber mats with Garcinia mangostana extracts. The agent has made the nanofiber mats antioxidant and antibacterial. According to the *in vivo* wound healing test results, acceleration in wound healing has been observed. They stated that these extract loaded chitosan nanofiber mats were good candidates for dermal healing mats [104]. Charernsriwilaiwat et al. were the ones trying to load lysozyme on chitosan nanofibers. Enhancement in wound healing was observed in *in vivo* tests of Male Wistar rats [105]. Vargas et al. revealed that electrospun hyperbranched polyglycerol nanofibers involving *Calendula officinalis* extract were suitable for wound dressing materials. They investigated the water vapor permeability, cytotoxicity, and skin irritation values and found that these nanofiber mats were proper for wound healing [106].

Silver nanoparticles are known as antimicrobial agents. Silver nanoparticles damage bacteria, virus, and fungi, and so they can be used as wound healing agents [107]. Nguyen et al. produced silver nanoparticles loaded with PVA electrospun fiber mats for wound healing. High antibacterial activities against *S. aureus* and *E. coli* were observed that is important for rapid wound healing [108]. Rujitanaroj et al. produced gelatin nanofibers with silver nanoparticles. They recommended that these mats can be used as dressing materials for burn wounds [109].

Epidermal growth factor added silk fibroin nanofiber mats were produced by electrospinning technique to improve wound healing process [110]. Also, Han et al. investigated nanofiber-based dressing from poly(3-hydroxybutyrate-co-3-hydroxyvalerate) polymer. They found that wound healing was supported at early stages by providing proper moist and mechanical support to the wound environment [111].

Poly(L-lactic acid) nanofibers were obtained in another study as skin tissue scaffolds. The success of the product was associated to the similarity of the structure of nanofibers to the extra cellular matrix of the skin [112].

6. Conclusions

Nanofibers have great potentials for use in cosmetics. Advances in fiber production technologies allow the design of new products for versatile usage in cosmetic applications. Increase in the awareness of usage of cosmetics in skin care as well as in therapy and healing directs researchers to conduct more research this field. It seems that nanofibers will continue to attract attention in this specific application area for many years.

Author details

Fulya Yilmaz[1], Gizem Celep[2] and Gamze Tetik[3*]

*Address all correspondence to: gamze.dogan@usak.edu.tr

1 Graduate School of Natural and Applied Sciences, Usak University, Usak, Turkey

2 Faculty of Engineering, Department of Textile Engineering, Usak University, Usak, Turkey

3 Faculty of Engineering, Department of Materials Science and Nanotechnology Engineering, Usak University, Usak, Turkey

References

[1] Chaudhri SK, Jain NK. History of cosmetics. Asian Journal of Pharmaceutics. 2009;3(3): 164–167. DOI: 10.4103/0973-8398.56292

[2] Yapar EA, Ölgen S. Cosmetics and in silico approaches. Journal of Marmara University Institute of Health Sciences. 2014;4(4):253–260. DOI: 10.5455/musbed.20140902023557

[3] Patil A, Sandewicz RW. Cosmetic science and polymer chemistry: Perfect together. In: Patil A, Ferritto MS, editors. Polymers for Personal Care and Cosmetics. New Jersey: Edison; 2013. p. 13–37. DOI: 10.1021/bk-2013-1148

[4] Chaudhri N, Soni GC, Prajapati SK. Nanotechnology: An advance tool for nanocosmetics preparation. International Journal of Pharma Research & Review. 2015;4(4): 28–40.

[5] Poletto FS, Beck RCR, Guterres SS, Pohlmann AR. Polymeric nanocapsules: Concepts and applications. In: Beck R, Guterres S, Pohlmann A, editors. Nanocosmetics and Nanomedicines. Berlin, Heidelberg: Springer; 2011. p. 49–68. DOI: 10.1007/978-3-642-19792-5

[6] Pokropivny V, Lohmus R, Hussainova I, Pokropivny A, Vlassov S. Introduction to Nanomaterials and Nanotechnology. Estonia: University of Tartu; 2007. 225 p. ISBN: 978–9949–311–741–3

[7] Patel RP, Joshi JR. An overview on nanoemulsion: A novel approach. International Journal of Pharmaceutical Sciences and Research. 2012;3(12):4640–4650. DOI: 10.13040/IJPSR.0975-8232.3(12).4640-50

[8] Bhat P, Mulgund S. Nanocosmetics: An evolving trend. Indo American Journal of Pharmaceutical Research. 2013;3(8):6549–6554. ISSN: 2231-6876

[9] EU-Scenhir. Opinion on the scientific basis for the definition of the term nanomaterial. 2010. http://ec.europa.eu/health/scientific_committees/emerging/docs/scenihr_o_032.pdf [Accessed: 2015-09-12].

[10] Ajazzuddin M, Jeswani G, Jha AK. Nanocosmetics: Past, present and future trends. Recent Patents on Nanomedicine. 2015;5:3–11. DOI: 10.2174/1877912305666150417232826

[11] Karjomaa S. Applications of nanotechnology: Cosmetics, course on health effects of engineered nanomaterials. In: Course on Health Effects of Engineered Nanomaterials. Finnish Institute of Occupational Health [Internet]. Available from: http://www.ttl.fi/partner/nanoturvallisuuskeskus/tapahtumat/Health_effects_on/Documents/Sari%20Karjomaa%20-%20Nanomaterials%20in%20cosmetics.pdf [Accessed: 2015-09-12]

[12] Patravale VB, Mandawgade SD. Novel cosmetic delivery systems: An application update. International Journal of Cosmetic Science. 2008;30:19–33. DOI: 10.1111/j.1468-2494.2008.00416.x

[13] Devarajan V, Ravichandran V. Nanoemulsions: As modified drug delivery tool. Pharmacie Globale International Journal of Comprehensive Pharmacy. 2011;4(2):1–6. ISSN: 0976-8157

[14] Thakur A, Walia MK, Hari Kumar SL. Nanoemulsion in enhancement of bioavailability of poorly soluble drugs: A review. Pharmacophore. 2013;4(1):15–25. ISSN: 2229-5402

[15] Sharma S, Sarangdevot K. Nanoemulsions for cosmetics. International Journal of Advanced Research in Pharmaceutical & Bio Sciences. 2012;2(3):408–415. ISSN: 2277-6222

[16] Özgün S. Nanoemulsions in cosmetics [Project Report]. Faculty of Engineering Materials Science and Engineering Department, Anadolu University; 2013.

[17] Sharma N, Bansal M, Visht S, Sharma PK, Kulkarni GT. Nanoemulsion: A new concept of delivery system. Chron Young Sci. 2010;1:2–6.

[18] Rajalakshmi R, Mahesh K, Ashok Kumar CK. A critical review on nano emulsions. International Journal of Innovative Drug Discovery. 2011;1(1):1–8.

[19] Kela SK, Kaur CD. Pharmaceutical nanoemulsions an ardent carrier for drug delivery. Indo American Journal of Pharmaceutical Research. 2013;3(11):9202–9212. ISSN: 2231-6876

[20] Nohynek GJ, Lademann J, Ribaud C, Roberts MS. Grey goo on the skin? Nanotechnology, cosmetic and sunscreen safety. Critical Reviews in Toxicology. 2007;37:251–277. DOI: 10.1080/10408440601177780

[21] Morganti P. Use and potential of nanotechnology in cosmetic dermatology. Clinical, Cosmetic and Investigational Dermatology. 2010;3:5–13.

[22] Lochhead RY. The role of polymers in cosmetics: Recent trends. In: Morgan SE, Havelka KO, Lochhead RY, editors. Cosmetic Nanotechnology. Washington DC: ACS

Symposium Series, American Chemical Society; 2007. p. 3–45. DOI: 10.1021/bk-2007-0961.ch001

[23] Sonneville-Aubrun O, Simonnet JT, L'Alloret F. Nanoemulsions: A new vehicle for skincare products. Advances in Colloid and Interface Science. 2004;108–109:145–149. DOI: 10.1016/j.cis.2003.10.026

[24] Shah P, Bhalodia D, Shelat P. Nanoemulsion: A pharmaceutical review. Systematic Reviews in Pharmacy. 2010;1(1):24–32. DOI: 10.4103/0975-8453.59509

[25] Abolmaali SS, Tamaddon AM, Farvadi FS, Daneshamuz S, Moghimi H. Pharmaceutical nanoemulsions and their potential topical and transdermal applications. Iranian Journal of Pharmaceutical Sciences. 2011;7(3):139–150.

[26] Zülli F, Belser E, Schmid D, Liechti C, Suter F. Preparation and properties of coenzyme Q10 nanoemulsions. 2006. [Internet] Available from: http://www.cosmeticsciencetechnology.com/articles/samples/1111.pdf [Accessed: 2016-06-25]

[27] Yechiel E, Coste RL. From ancient potions to modern lotions. A technology overview and introduction to topical delivery systems. In: Rosen MR, editor. Delivery System Handbook for Personal Care and Cosmetic Products, Technology, Applications, and Formulations. Norwich: William Andrew Publishing; 2005. 130 p. ISBN: 978-0-8155-1504-3

[28] Mozafari MR, Johnson C, Hatziantoniou S, Demetzos C. Nanoliposomes and their applications in food nanotechnology. Journal of Liposome Research. 2008;18(4):309–327. DOI: 10.1080/08982100802465941

[29] Khosravi-Darani K, Mozafari MR. Nanoliposome potentials in nanotherapy: A concise overview. International Journal of Nanoscience and Nanotechnology. 2010;6(1):3–13.

[30] Kundu AK, Hazari S, Chinta DD, Pramar YV, Dash S, Mandal TK. Development of nanosomes using high-pressure homogenization for gene therapy. Journal of Pharmacy and Pharmacology. 2010;62:1103–1111. DOI: 10.1111/j.2042-7158.2010.01140.x

[31] Mozafari MR. Nanoliposomes: Preparation and analysis. In: Weissig V, editor. Liposomes, Methods and Protocols. United States: Humana Press, Methods in Molecular Biology 605; 34 2009. 29–50 p. Online ISBN: 978-1-60327-360-2

[32] Challenge Biomedical Science – Nanotechnology [Internet]. Available from: http://www.w5online.co.uk/education/teachers-toolbox/nanotechnology/?download=1&resource=21 [Accessed: 2016-03-01].

[33] Lohani A, Verma A, Joshi H, Yadav N, Karki N. Review article: Nanotechnology-based cosmeceuticals. ISRN Dermatology. 2014;Article ID 843687:1–14. DOI: 10.1155/2014/843687

[34] Wang H. Development of silica-based nanopigments [Thesis]. Aachen University; 2012.

[35] Nejad FM, Basghshahi S, Bakhtiari L. Advantages of nano pigments over micro pigments in obtaining larger spectra of colours in CMYK system. Transactions of the Indian Ceramic Society. 2011;70(2):93–99. DOI: 10.1080/0371750X.2011.10600154

[36] Ahmed IS, Shama SA, Dessouki HA, Ali AA. Synthesis, thermal and spectral characterization of nanosized $Ni_xMg_{1-x}Al_2O_4$ powders as new ceramic pigments via combustion route using 3-methylpyrozole-5-one as fuel. Spectrochimica Acta Part A: Molecular and Biomolecular Spectroscopy. 2011;81:324–333. DOI: 10.1016/j.saa.2011.06.019

[37] Loreal 2006 Sustainable Development Report [Internet]. 2006. Available from: http://aeca.es/old/comisiones/rsc/biblioteca_memorias_rsc/publicadas_07/loreal_2006.pdf [Accessed: 2016-02-25].

[38] Bhat P, Mulgund S. Nanocosmetics: An evolving trend. Indo American Journal of Pharmaceutical Research. 2013;3(8):6549–6554. ISSN: 2231-6876

[39] Pal TK, Mondal O. Prospect of nanotechnology in cosmetics: Benefit and risk assessment. World Journal of Pharmaceutical Research. 2014;3(2):1909–1919. ISSN: 2277-7105

[40] Shaista A. Applications of nano-pigments. [Internet]. 2015. Available from: http://pcs online.pk/pages/Presentation-2015.html [Accessed: 2016-02-05]

[41] Houck MM. Identification of Textile Fibers, 1st ed. Chichester: Woodhead Publishing; 2009. 396 p. ISBN: 9781845695651

[42] Rocco MC, William RS, Alivisiatos P. Nanotechnology Research Directions: IWGN Workshop Report, National Science and Technology Council, September 1999.

[43] Frenot A, Chronakis IS. Polymer nanofibers assembled by electrospinning. Current Opinion in Colloid and Interface Science. 2003;8:64–75. DOI: 10.1016/S1359-0294(03)00004-9

[44] Ko FI. Nanofiber technology: Bridging the gap between nano and macro world, in NATO ASI. In: Guceri S, Gogotsi YA, editors. Nanoengineered Nanofibrous Materials. Chichester: Kluwer Academic Publishers; 2004. 18 p. DOI: 10.1007/978-1-4020-2550-1_1

[45] Huang ZM, Zhang YZ, Kotaki M, Ramakrishna S. A review on polymer nanofibers by electrospinning and their applications in nanocomposites. Composite Science and Technology. 2003;63:2223–2253. DOI: 10.1016/S0266-3538(03)00178-7

[46] Graham K, Schreuder-Gibson H, Gogins M. Incorporation of electrospun nanofibers into functional structures. Technical Association of the Pulp & Paper Industry, September 15–18, Baltimore, MD, 2003. 16 p.

[47] Subbiah T, Bhat GS, Tock RW, Parameswaran S, Ramkumar SS. Electrospinning of nanofibers. Journal of Applied Polymer Science. 2005;96:557–569. DOI: 10.1002/app.21481

[48] Huang C, Wang S, Zhang H, Li T, Chen S, Lai C, Hou H. High strength electrospun polymer nanofibers made from BPDA–PDA polyimide. European Polymer Journal. 2006;42:1099–1104. DOI: 10.1016/j.eurpolymj.2005.11.005

[49] Dreno B, Araviiskaia E, Berardesca E, Bieber T, Hawk J, Sanchez-Viera M, Wolkenstein P. The science of dermocosmetics and its role in dermatology. Journal of the European Academy of Dermatology and Venereology. 2014;28(11):1409–1417. DOI: 10.1111/jdv.12497

[50] Agarwal S, Wendorff JH, Greiner A. Use of electrospinning technique for biomedical applications. Polymer. 2008;49:5603–5621. DOI: 10.1016/j.polymer.2008.09.014

[51] Doğan G. An investigation of potential applications of espun biopolymer nanofibers in tissue engineering and drug release [Thesis]. Ege University; 2012. 184 p.

[52] Zhou FL, Gong RH. Manufacturing technologies of polymeric nanofibres and nanofibre yarns. Polymer International. 2008;57:837–845. DOI: 10.1002/pi.2395

[53] Cloutier S, Manuel LM. Flash spinning process for forming strong discontinuous fibres. US Patent 5415818; 1995.

[54] Zhang X, Baughman CB, Kaplan DL. In vitro evaluation of electrospun silk fibroin scaffolds for vascular cell growth. Biomaterials. 2008;29:2217–2227. DOI: 10.1016/j.biomaterials.2008.01.022

[55] Palmer LC, Stupp IS. Molecular self-assembly into one-dimensional nanostructures. Accounts of Chemical Research. 2008;41(12):1674–1684. DOI: 10.1021/ar8000926

[56] Ramakrishna S, Fujihara K, Teo WE, Lim TC, Ma Z. An Introduction to Electrospinning and Nanofibers. Chichester: World Scientific Publishing; 2005. 382 p. ISBN: 9812564152, 9789812564153

[57] Liebmann B, Hümmerich D, Scheibel T, Fehr M. Formulation of poorly water-soluble substances using self-assembling spider silk protein. Colloids and Surfaces A: Physicochemical and Engineering Aspects. 2008;331(1–2):126–132. DOI: 10.1016/j.colsurfa.2008.04.005

[58] Ondarçuhu T, Joachim C. Drawing a single nanofibre over hundreds of microns. Europhysics Letters. 1998;42(2):215–220. DOI: 10.1209/epl/i1998-00233-9

[59] Nain AS, Wong JC, Amon C, Sitti M. Drawing suspended polymer micro/nanofibers using glass micropipettes. Appeared in Applied Physics Letters. 2006;89(18):183105–183107. DOI: 10.1063/1.2372694

[60] Bajáková J, Chaloupek J, Lukáš D, Lacarin M. Drawing—The Production of Individual Nanofibers by Experimental Method, 3rd International Conference, Brno, Czech Republic; 21–23 September 2011.

[61] Chen T, Li L, Huang X. Predicting the fibre diameter of meltblown nonwovens: Comparison of physical, statistical and artificial neural network models. Modelling and

Simulation in Materials Science and Engineering. 2005;13:575–584. DOI: 10.1088/0965-0393/13/4/008

[62] Ellison CJ, Phatak A, Giles DW, Macosko CW, Bates FS. Meltblown nanofibers: Fiber diameter distributions and onset of fiber breakup. Polymer. 2007;48:3306–3316. DOI: 10.1016/j.polymer.2007.04.005

[63] Hills Inc. 2016. http://hillsinc.net/assets/pdfs/melt-blown.pdf [Accessed: 2016-02-01].

[64] Hassan MA, Yeom BY, Wilkie A, Purdeyhimi B, Khan SA. Fabrication of nanofiber meltblown membranes and their filtration properties. Journal of Membrane Science. 2013;427:336–344. DOI: 10.1016/j.memsci.2012.09.050

[65] Bhat G. Meltblown submicron fibers for filter media and other applications. International Fiber Journal. 2015. April/ May 2015. [Internet]. 2015. Available from: http://www.fiberjournal.com/featured-articles/meltblown-submicron-fibers-for-filter-media-and-other-applications/ [Accessed: 2016-02-01].

[66] Amundson JD, Abuto FP, Clare TP, Foley EJ, Winder EM, Schmidt PR, Joseph WR, McDonald DL. Wet wipes dispensing system. US Patent 7850041B2; 2010.

[67] Vogt CM, Cohen B, Ellis CJ. Multiple application meltblown nonwoven wetwipe and method. US Patent 5656361 A; 1997.

[68] Choi BK, Lee JH. Mask pack. US Patent 20140352031 A1; 2014.

[69] Huczko A. Template-based synthesis of nanomaterials. Applied Physics A. 2000;70:365–376. DOI: 10.1007/s003390000440

[70] Tao SL, Desai TA. Aligned arrays of biodegradable poly(epsilon-caprolactone) nanowires and nanofibers by template synthesis. Nano Letters. 2007;7(6):1463–1468. DOI: 10.1021/nl0700346

[71] Smith LA, Ma PX. Nano-fibrouss caffolds for tissue engineering. Colloids and Surfaces B: Biointerfaces. 2004;39:125–131. DOI: 10.1016/j.colsurfb.2003.12.004

[72] Liu X, Ma PX. Phase separation, pore structure, and properties of nanofibrous gelatin scaffolds. Biomaterials. 2009;30:4094–4103. DOI: 10.1016/j.biomaterials.2009.04.024

[73] Fisher WB, Kim DW, Norwood OW, Showers JW, Swanson EA. Gene Clyde Weedon. Apparatus for melt-spinning synthetic fibers. US Patent 3936253 A; 1977.

[74] Kikutani T, Radhakrishnan J, Arikawa S, Takaku A, Okui N, Jin X, Niwa F, Kudo Y. High-speed melt spinning of bicomponent fibers: Mechanism of fiber structure development in poly(ethy1ene terephthalate)/polypropylene system. Applied Polymer Science. 1996;62:1913–1924. DOI: 10.1002/(SICI)1097-4628(19961212)62:11<1913::AID-APP16>3.0.CO;2-Z

[75] Zhang X. Fundamentals of Fiber Science. Chichester: DEStech Publications; 2014. 426 p. ISBN: 1605951196, 9781605951195

[76] Zhang X, Lu Y. Centrifugal spinning: An alternative approach to fabricate nanofibers at high speed and low cost. Polymer Reviews. 2014;54(4):677–701. DOI: 10.1080/15583724.2014.935858

[77] Weitz RT, Harnau L, Rauschenbach S, Burghard M, Kern K. Polymer nanofibers via nozzle-free centrifugal spinning. Nano Letters. 2008;8(4):1187–1191. DOI: 10.1021/nl080124q

[78] Ren L, Pandit V, Elkin J, Denman T, Cooper JA, Kotha SP. Large-scale and highly efficient synthesis of micro- and nano-fibers with controlled fiber morphology by centrifugal jet spinning for tissue regeneration. Nanoscale. 2013;5:2337–2345. DOI: 0.1039/C3NR33423F

[79] Lang SB. The history of pyroelectricity: From ancient Greece to space missions. Ferroelectrics. 1999;230(1):99–108. DOI: 10.1080/00150199908214903

[80] Zhang S. Mechanical and physical properties of electrospun nanofibers [Thesis]. North Carolina State University; 2009. 72 p.

[81] Pumfrey S. I William Gilbert, 6-20. In: Harman P, Mitton S, editors. Cambridge Scientific Minds, 1st ed. Cambridge: Cambridge University Press; 2002. 352 pp.

[82] Tripatanasuwan S, Zhong Z, Reneker DH. Effect of evaporation and solidification of the charged jet in electrospinning of poly(ethylene oxide) aqueous solution. Polymer. 2007;48:5742–5746. DOI: 10.1016/j.polymer.2007.07.045

[83] Doshi J, Reneker DH. Electrospinning process and applications of electrospun fibers. Journal of Electrostatics. 1995;35(2–3):151–160. DOI: 10.1016/0304-3886(95)00041-8

[84] Zeng J. Meso- and nano- scaled polymer fibers and tubes, fabrication, functionalization and characterization [Thesis]. Fachbereich Chemie der Philipps-Universität Marburg; 2003. p. 7–9.

[85] Teo WE, Ramakrishna S. A review on electrospinning design and nanofibre assemblies. Nanotechnology. 2006;17:89–106. DOI: 10.1088/0957-4484/17/14/R01

[86] Reneker DH, Yarin AL. Electrospinning jets and polymer nanofibers. Polymer. 2008;49:2387–2425. DOI: 10.1016/j.polymer.2008.02.002

[87] Lukáš D, Sarkar A, Martinová L, Vodsed'álková K, Lubasová D, Chaloupek J, Pokorný P, Mikeš P, Chvojka J, Komárek M. Physical principles of electrospinning (electrospinning as a nano-scale technology of the twenty-first century). Textile Progress. 2009;41(2):59–140. DOI: 10.1080/00405160902904641

[88] Rangkupan R, Reneker DH. Electrospinning process of molten polypropylene in vacuum. Journal of Metals, Materials and Minerals. 2003;12(2):81–87.

[89] Mit-uppatham C, Nithitanakul M, Supaphol P. Ultrafine electrospun polyamide-6 fibers: Effect of solution conditions on morphology and average fiber diameter.

Macromolecular Chemistry and Physics. 2004;205:2327–2338. DOI: 10.1002/macp.200400225

[90] Wannatong L, Sirivat A, Supaphol P. Effects of solvents on electrospun polymeric fibers: Preliminary study on polystyrene. Polymer International. 2004;53:1851–1859. DOI: 10.1002/pi.1599

[91] Reneker DH, Yarin AL, Fong H, Koombhongse S. Bending instability of electrically charged liquid jets of polymer solutions in electrospinning. Journal of Applied Physics. 2000;87(9):4531–4547. DOI: 10.1063/1.373532

[92] Shin Y, Hohman M, Brenner M, Rutledge G. Electrospinning: A whipping fluid jet generates submicron polymer fiber. Applied Physics Letters. 2001;78(8):1149–1151. DOI: 10.1063/1.1345798

[93] Bhardwaj N, Kundu SC. Electrospinning: A fascinating fiber fabrication technique. Biotechnology Advances. 2010;28:325–347. DOI: 10.1016/j.biotechadv.2010.01.004

[94] Han T, Reneker DH, Yarin AL. Buckling of jets in electrospinning. Polymer. 2007;48:6064–6076. DOI: 10.1016/j.polymer.2007.08.002

[95] Ramakrishna S, Fujihara K, Teo WE, Yong T, Ma Z, Ramaseshan R. Electrospun nanofibers: Solving global issues. Materials Today. 2006;9(3):40–50. DOI: 10.1016/S1369-7021(06)71389-X

[96] De Vrieze S, De Clerk K. 80 Years of Electrospinning. 5 p. 2009; https://biblio.ugent.be/publication/767594 [Accessed: 2016-02-01].

[97] Khayet M, Matsuura T. Membrane Distillation: Principles and Applications. Amsterdam: Elsevier; 2011. p. iii. DOI: 10.1016/B978-0-444-53126-1.10017-X

[98] Fathi-Azarbayjani A, Qun L, Chan YW, Chan SY. Novel vitamin and gold-loaded nanofiber facial mask for topical delivery. AAPS PharmSciTech. 2010;11(3):1164–1170. DOI: 10.1208/s12249-010-9475-z

[99] Smith D, Reneker D, Kataphinan W, Dabney S. Electrospun skin masks and uses thereof. European Patent EP 1 221 927 B1; 2002.

[100] Kim C. Cosmetic sheet formed from nanofiber with controlled dissolution velocity and method of manufacturing the same. US Patent 20150272855; 2015.

[101] Taepaiboon P, Rungsardthong U, Supaphol P. Vitamin-loaded electrospun cellulose acetate nanofiber mats as transdermal and dermal therapeutic agents of vitamin A acid and vitamin E. European Journal of Pharmaceutics and Biopharmaceutics. 2007;67(2):387–397. DOI: 10.1016/j.ejpb.2007.03.018

[102] Sheng X, Fan L, He C, Zhang K, Mo X, Wang H. Vitamin E-loaded silk fibroin nanofibrous mats fabricated by green process for skin care application. International Journal of Biological Macromolecules. 2013;56:49–56. DOI: 10.1016/j.ijbiomac.2013.01.029

[103] Zanin MHA, Cerize NNP, Oliveira AM. Production of nanofibers by electrospinning technology: Overview and application in cosmetics. In: Beck R, Guterres S, Pohlmann A, editors. Nanocosmetics and Nanomedicines New Approaches for Skin Care. Chichester: Springer; 2011. pp. 311–332. DOI: 10.1007/978-3-642-19792-5

[104] Charernsriwilaiwat N, Rojanarata T, Ngawhirunpat T, Sukma M, Opanasopit P. Electrospun chitosan-based nanofiber mats loaded with Garcinia mangostana extracts. International Journal of Pharmaceutics. 2013;452(1–2):333–343. DOI: 10.1016/j.ijpharm.2013.05.012

[105] Charernsriwilaiwat N, Opanasopit P, Rojanarata T, Ngawhirunpat T. Lysozyme-loaded, electrospun chitosan-based nanofiber mats for wound healing. International Journal of Pharmaceutics. 2012;427(2):379–384. DOI: 10.1016/j.ijpharm.2012.02.010

[106] Torres Vargas EA, Vale Baracho NC, Brito J, Queiroz AAA. Hyperbranched polyglycerol electrospun nanofibers for wound dressing applications. Acta Biomaterialia. 2010;6:1069–1078. DOI: 10.1016/j.actbio.2009.09.018

[107] Sridhar R, Sundarrajan S, Venugopar JV, Ravichandran R, Ramakrishna S. Electrospun inorganic and polymer composite nanofíbers for biomédical applications. Journal of Biomaterials Science, Polymer Edition. 2013;4(24):365–385. DOI: 10.1080/09205063.2012.690711

[108] Nguyen TH, Kim YH, Song HY, Lee BT. Nano Ag loaded PVA nano-fibrous mats for skin applications. Journal of Biomedical Materials Research B: Applied Biomaterials. 2011;96b:225–233. DOI: 10.1002/jbm.b.31756

[109] Rujitanaroj P, Pimpha N, Supaphol P. Wound-dressing materials with antibacterial activity from electrospun gelatin fiber mats containing silver nanoparticles. Polymer. 2009;49:4723–4732. DOI: 10.1016/j.polymer.2008.08.021

[110] Schneider A, Wang XY, Kaplan DL, Garlick JA, Egles C. Biofunctionalized electrospun silk mats as a topical bioactive dressing for accelerated wound healing. Acta Biomaterialia. 2009;5(7):2570–2578. DOI: 10.1016/j.actbio.2008.12.013

[111] Han I, Shim KJ, Kim JY, Im SU, Sung YK, Kim M, Kang IK, Kim JC. Effect of poly(3-hydroxybutyrate-*co*-3-hydroxyvalerate) nanofiber matrices cocultured with hair follicular epithelial and dermal cells for biological wound dressing. Artificial Organs. 2007;31:801–808. DOI: 10.1111/j.1525-1594.2007.00466.x

[112] Mohiti-Asli M, Pourdeyhimi B, Loboa EG. Novel, silver-ion-releasing nanofibrous scaffolds exhibit excellent antibacterial efficacy without the use of silver nanoparticles. Acta Biomaterialia. 2014;10:2096–2104. DOI: 10.1016/j.actbio.2013.12.024

Section 4

State-of-the-Art Nanocomposites

Chapter 8

Synthesis and Characteristics of Carbon Nanofibers/ Silicon Composites and Application to Anode Materials of Li Secondary Batteries

Chang-Seop Lee and Yura Hyun

Additional information is available at the end of the chapter

http://dx.doi.org/10.5772/63507

Abstract

Among the various synthesizing technologies of carbon nanofibers (CNFs), chemical vapor deposition (CVD) technology, which uses hydrocarbon gas or carbon monoxide as a carbon source gas and pyrolyzes it to grow CNFs on transition metal catalysts, such as Ni, Fe, and Co, has been regarded as the most inexpensive and convenient method to produce CNFs for industrial use. Experimental variables for CVD are source gas, catalyst layers, temperature, and reaction time. Since the particle size of metal catalysts has an influence on the diameter of CNFs, it is possible to control the diameter of CNFs by varying particle sizes of the metal. As such, it is possible to synthesize CNFs selectively through the selective deposition of catalyst metals. In this study, CNFs were grown by CVD on C-fiber textiles, which had catalysts deposited via electrophoretic deposition. The CNFs were coated with a silica layer via hydrolysis of TEOS (tetraethyl orthosilicate), and the CNFs were oxidized by nitric acid. Due to oxidation, a hydroxyl group was created on the CNFs, which was then able to be used as an activation site for the SiO_2. CNFs and the CNFs/SiO_2 composite can be used in various applications, such as a composite material, electromagnetic wave shielding material, ultrathin display devices, carbon semiconductors, and anode materials of Li secondary batteries. In particular, there is an increasing demand for lightweight, small-scale, and high-capacity batteries for portable electronic devices, such as laptop computers or smart phones, along with the escalating concern of fossil energy depletion. Accordingly, CNFs and CNFs/SiO_2 composites are receiving attention for their use as anode materials of Li secondary batteries, which are eco-friendly, lightweight, and high capacity. Therefore, the physicochemical properties and electrochemical performance data of synthesized CNFs and CNFs/SiO_2 composite are described in this chapter.

Keywords: carbon nanofibers, chemical vapour deposition, silicon composites, anode materials, Li secondary batteries

1. Introduction

Carbon has various types of allotrope (graphite, diamond, carbon nanofibers (CNFs), carbon nanotubes, graphene, etc.) depending on the molecular bonding mode with hybridized bonding of sp, sp^2, and sp^3. **Figure 1** shows a schematic of carbon fibers and tubes of various sizes.

CNFs among various allotropes of carbon are fibrous carbon materials with less than 1 μm thickness and over 90% carbon content and can take various forms, such as herringbone, platelet, and spiral. Such CNFs can be used in various applications, such as a composite material, electromagnetic wave shielding material, ultrathin display devices, carbon semiconductors, and anode materials of secondary batteries. In particular, there is an increasing demand for lightweight, small-scale, and high-capacity batteries for portable electronic devices, such as laptop computers or smart phones, along with the recent concern of fossil energy depletion. Accordingly, CNFs are receiving attention for their potential use as anode materials of Li secondary batteries, which are eco-friendly, lightweight, and high capacity. Secondary batteries can be charged and discharged several times for reuse. Lead and Ni-Cd batteries have been most commonly used prior to the introduction of Li secondary batteries, but they harbor the disadvantages of a memory effect and environmental pollution issues.

Recently, the development of Li secondary batteries received more attention, since they are environmentally safe, with lightweight, small-scale, and high-capacity features required for small IT devices, such as laptop computers or mobile phones [1–17].

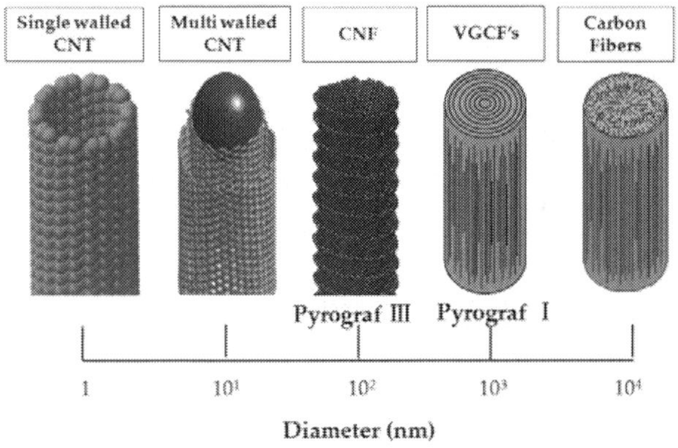

Figure 1. Schematic of carbon fibers and tubes of various types.

1.1. Lithium secondary batteries

Li secondary batteries mainly consist of a cathode, anode, electrolyte, and separator. The anode reversibly intercalates and disintercalates the lithium ions coming from the cathode to allow electrons to flow through the internal circuit and generate electricity. **Figure 2** shows a schematic diagram showing the principle behind the Li secondary battery.

Studies on the performance enhancement of lithium secondary batteries are mostly concerning the performance improvement of anode materials, including carbon materials and the development of new materials. Since the charging and discharging performance of Li secondary batteries is significantly influenced by the structure of the anode materials for intercalating lithium ions, this study focused on the performance improvement and development of carbon materials as anode materials to allow more Li ions to be inserted. For such anodes, graphite, first developed by Bell Labs in 1981, has been used in most cases. Graphite has the advantage of its crystal structure during charging and discharging, so that its volume does not significantly change; however, it limits the performance of the Li secondary batteries, due to its small maximum theoretical capacity of 372 mAh g^{-1}.

Accordingly, studies are actively conducted to test new carbon-based materials that would allow high-capacity and performance improvement of Li secondary batteries. The CNFs of carbon-based anode materials do not change significantly in their crystal structure during the intercalation and disintercalation process of lithium ions, as in the case of graphite. They are accordingly receiving attention as a new anode material that would allow lithium secondary batteries to provide an excellent lifetime [18–20].

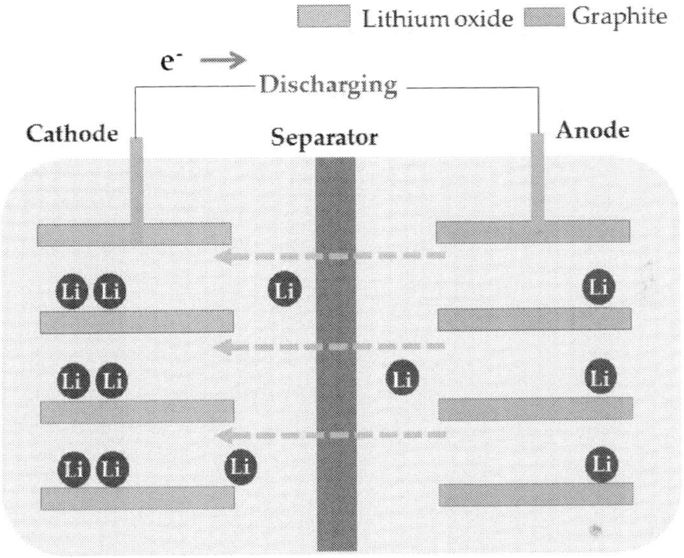

$$Li_{1-X}CoO_2 + Li_XC_6 \rightarrow C_6 + LiCoO_2$$

Figure 2. Schematic diagram of the principle of the Li secondary battery.

1.2. CNFs/silicon composites

Silicon is suitable as a high-capacity anode material because of its high maximum theoretical capacity of 4200 mAh g^{-1}; however it is associated with a deteriorating electrical connection due to cracks forming as a result of the nearly 400% volume change during charging and discharging. It also has the problem of deteriorating charging and discharging characteristics by generating a large irreversible capacity. Various studies have been attempted to solve the problems of CNFs and silicon, and a method of synthesizing CNFs/Si composites was presented as one of the solutions [19].

Since CNFs provide a flexible space for the volume expansion of silicon in the process of synthesizing CNFs/Si composite, it has the potential to be a good candidate and as such has been selected as an anode material of Li secondary batteries in order to enhance its performance for use in this study. **Figure 3** shows the grown CNFs on a Si surface which play a role as shrinked springs when the composites are charged.

CNFs were synthesized by using chemical vapor deposition (CVD), and the effects of synthesis conditions, such as synthesis temperature and the concentration ratio of catalysts on the growth of CNFs, were studied. CNFs/Si composites were synthesized based on these CNFs, and the physiochemical characteristics of CNFs/Si composites and the electrochemical characteristics as anode materials of Li secondary batteries were also investigated [9, 18–22].

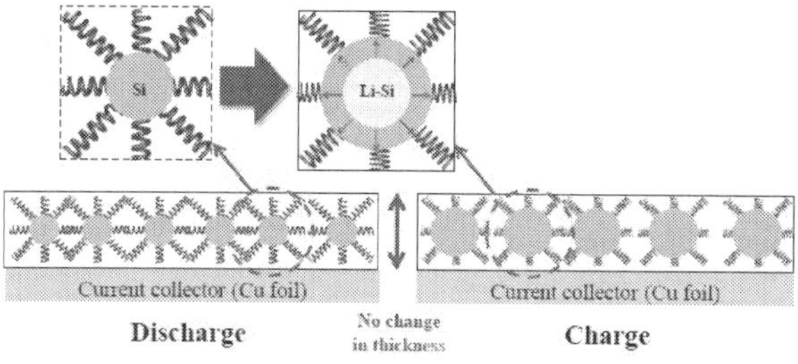

Figure 3. Schematic diagram showing the buffering principle of CNFs/Si composites.

2. Synthesis of CNFs on electroplated Fe/C-fiber textiles

2.1. Experimental process

2.1.1. Oxidation

C-fiber textiles (SGL Carbon Group) were put into 60% nitric acid and heated at 80°C for 30 min under reflux to form OH groups on the surface of C-fiber textiles. After the completion of oxidation, the oxidized C-fiber textiles were cleaned and stored in distilled water, without drying in an oven, in order to keep the OH groups attached to the surface [4, 7, 17].

2.1.2. Deposition of Fe catalyst

For the deposition of a Fe catalyst, $Fe(NO_3)_3 \bullet 9H_2O$ as a precursor was first dissolved in distilled water, and then the C-fiber textiles were immersed in the solution. The pH of the solution was lowered to 1.00 with nitric acid, and the solution was heated to 90°C. The ratio of C-fiber textiles and the amount of Fe in the solution was controlled as an experimental variable. Afterward, hydrolysis was carried out by slow dropping of 1 M urea solution.

It is known that $Fe(OH)_3$ is formed by deposition-precipitation on the surface of oxidized C-fiber textiles, through the reaction of OH groups. For comparison, deposition of ferrous particles on the oxidized C-fiber textiles was carried out, by dipping the substrate into the solution, after the hydrolysis by urea was complete. The specimen was cleaned by distilled water, after completion of the deposition process, and dried for 24 hours at 80°C [4, 7, 17].

2.1.3. Reduction

A reduction step was applied in order to convert iron hydroxides and oxides attached to the surface of C-fiber textiles into elemental iron using a tube furnace. Hydrogen balanced with nitrogen gas was used for the reduction process, and the flux of the reaction gas was controlled by an electronic mass flow controller (MFC). The reactor temperature was increased by 10°C/min up to 600°C in pure N_2 atmosphere. Once the temperature reached 600°C, N_2 gas balanced with 20% H_2 gas was flowed into the reactor for 3 hours to carry out the reduction process [4, 7, 17].

2.1.4. Growth of CNFs

CNFs were grown on C-fiber textiles by CVD in a horizontal quartz tube reaction apparatus after completion of the reduction process. The flux of reaction gas was controlled by an electronic MFC, and ethylene gas (C_2H_4) was used to grow the CNFs as a carbon source. Hydrogen gas was used to eliminate the remaining hydroxyl group after the reduction process, while nitrogen was used to stabilize the reaction.

The prepared metal catalyst was evenly spread on a quartz boat, which was placed into the reactor under nitrogen atmosphere, and the reactor temperature was increased by 10°C/min. Once the temperature reached 700°C, it was maintained for 30 min. N_2 gas balanced with 20% H_2 gas was flowed into the reactor, and then 20% ethylene balanced with N_2 gas was flowed into the reactor for 5 hours. After the completion of the reaction, the temperature was lowered to room temperature in N_2 atmosphere [4, 7, 17].

2.2. Physicochemical properties of the grown CNFs

2.2.1. Scanning Electron Microscope (SEM)

Figures 4 and **5** show the shape of the CNFs grown on the C-fiber textiles by the CVD method, as taken by SEM, with 100,000× magnification. It was found that the diameters of CNFs increased with decreasing a Fe/C ratio. The diameters of CNFs grew up to 40–60 nm when the weight ratio of C-fiber textiles to Fe was 1:30, whereas CNFs grew up to 30–55 nm in diameter when the ratio was 1:70.

In contrast, it was observed that CNFs scarcely grew in the samples prepared by the dip-coating method if the ratio of C-fiber textiles to Fe reached 1:70; instead they started to cohere during the growth process. It is assumed that too many Fe catalyst particles were present on the surface of the C-fiber textiles, which interferes with the growth of CNFs. CNFs of 40–60 nm in diameter grew when the ratio was 1:10, while the thinnest and the most even CNFs, of 25–30 nm in diameter, grew when the ratio was 1:30 [4, 7, 17].

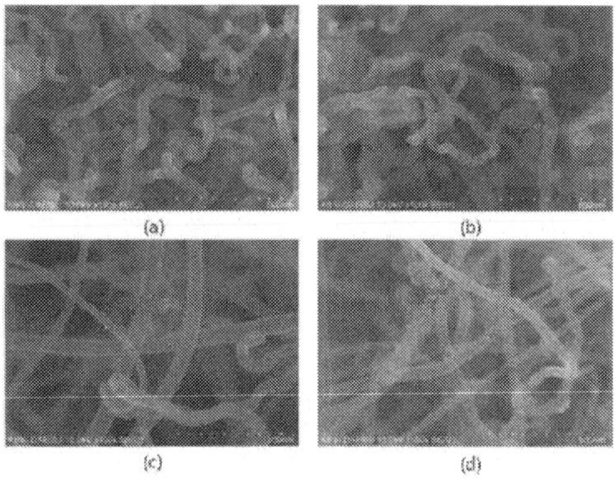

Figure 4. CNFs grown on C-fiber textiles, under the CVD method. (a) 1 g C-fiber textile, 10 g Fe(III) weight ratio; (b) 1 g C-fiber textile, 30 g Fe(III) weight ratio; (c) 1 g C-fiber textile, 50 g Fe(III) weight ratio; and (d) 1 g C-fiber textile, 70 g Fe(III) weight ratio.

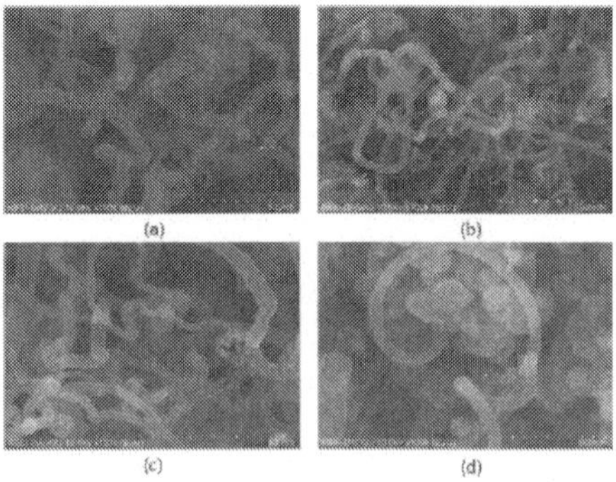

Figure 5. Shapes of CNFs, after the CVD process, when the dip-coating time was 30 min. (a) 1 g C-fiber textile, 10 g Fe(III) weight ratio; (b) 1 g C-fiber textile, 30 g Fe(III) weight ratio; (c) 1 g C-fiber textile, 50 g Fe(III) weight ratio; and (d) 1 g C-fiber textile, 70 g Fe(III) weight ratio.

http://dx.doi.org/10.5772/63507

2.2.2. *X-ray Diffraction (XRD)*

XRD patterns were taken in order to analyze how the crystalline structure of the samples change according to the different processes, and the results are shown in **Figure 6**. After the deposition process, iron oxide phases of different stoichiometry were found. It is known that $Fe(OH)_3$ present on the sample surface after precipitation-deposition transforms into iron oxide during the drying process, and then Fe particles of cubic crystalline structure remained after reduction.

The graphite peaks in the raw material mainly belong to the graphite flakes, which are present in the C-fiber textile. The peak broadening, which can be observed for the samples after CVD, can be attributed to the presence of CNFs [4, 7, 17].

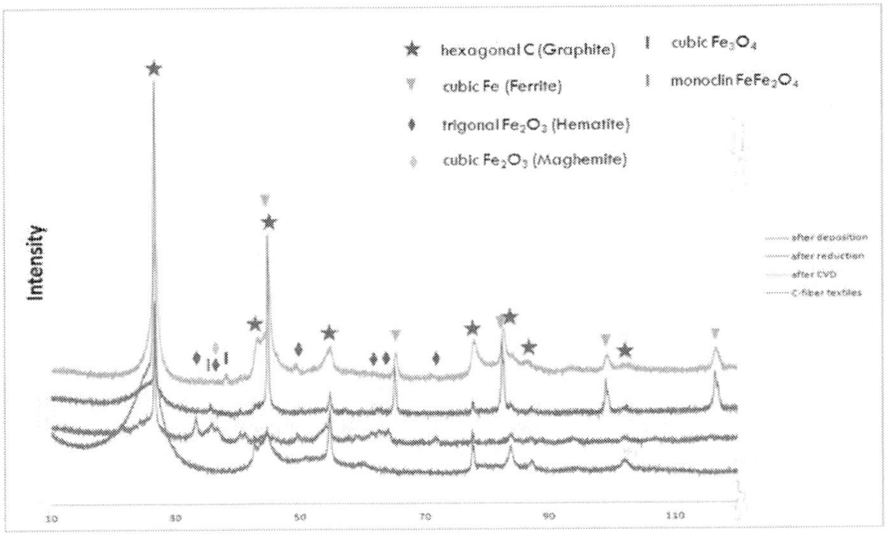

Figure 6. XRD pattern of sample according to changes in the experimental process.

2.2.3. *X-ray Photoelectron Spectroscopy (XPS)*

XPS analysis was carried out in order to analyze the binding energies of C, Fe, and oxygen. Measurements were performed for the sample where the CNFs (C:Fe 1:30, 30 min dip coating) were grown as uniformly as possible. **Figure 7** shows the XPS analysis result after the growth of CNFs. The binding energy of the Fe catalyst particles deposited to CNFs, which were grown on the surface, did not appear on the XPS spectrum, while the binding energies of C and oxygen appeared on the spectra [4, 7, 17].

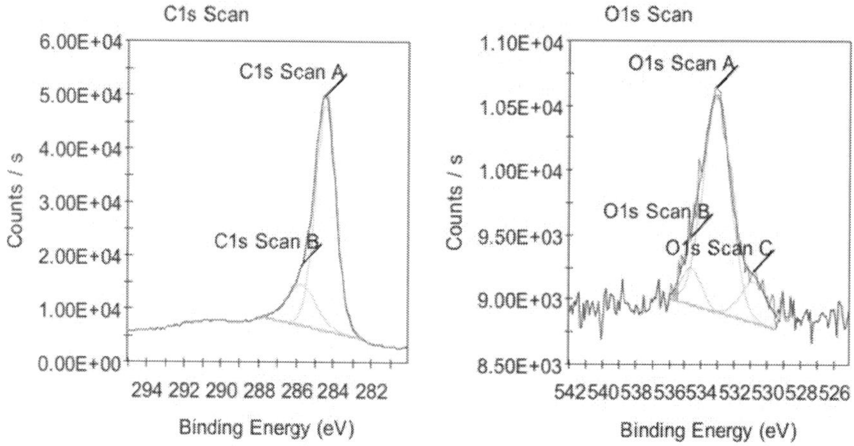

Name	Binding Energy(eV)	Analysis
C1s Scan A	284.46	Graphite C
C1s Scan B	285.7	C-H combination
O1s Scan A	534.63	C-O combination
O1s Scan B	533.13	C=O combination
O1s Scan C	530.81	Fe_2O_3 combination

Figure 7. XPS results of CNFs grown on C-fiber textiles.

2.2.4. Thermogravimetry/Differential Thermal Analysis (TG/DTA)

Figure 8 shows TG/DTA thermograms for the CNF grown samples. After the Fe catalyst was deposited (a) by the deposition-precipitation method or (b) by the dip-coating method, with the same C:Fe ratio of 1:30, analysis was performed on the samples in which CNFs were grown. The TG/DTA curves showed similar tendencies when the experiment was conducted by both methods.

As can be seen in **Figure 8**, the decrease of TGA thermogram ranges from 490°C to 750°C, which resulted from the oxidation of CNFs and C-fiber textiles composing the sample. As for the DTA thermogram, exothermic peaks were found in both samples near 600°C, 690°C, and 850°C. The exothermic peak near 600°C appearing at the beginning of the DTA thermogram is deemed as oxidation reaction to carbon of CNFs; and meantime, exothermic peaks appearing in higher temperature of near 690°C and 850°C are deemed as peaks showing the oxidation reaction of C-fiber and graphite composing C-fiber textiles, which were used as substrate [4, 7, 17].

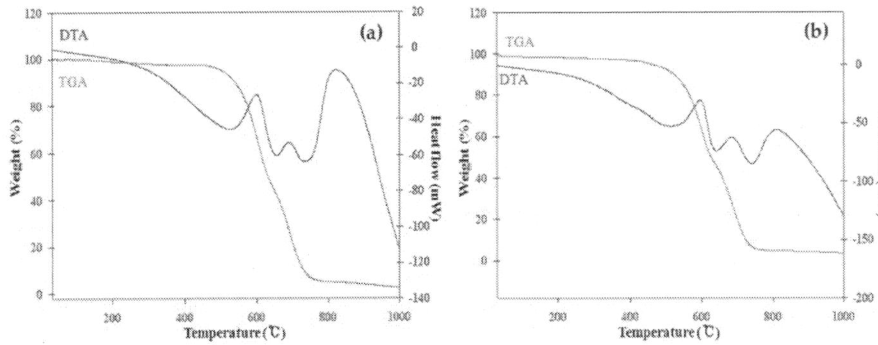

Figure 8. TG/DTA thermograms of CNFs grown on C-fiber textiles by the two different deposition methods: (a) deposition-precipitation method and (b) dip-coating method.

2.2.5. *Raman*

After CNFs had been synthesized in the CVD method, Raman analysis was performed in order to investigate the characteristics of C— bonding, and the results are presented in **Figure 9**. After Fe catalysts were deposited on CNFs, they were grown on C-fiber textiles by the deposition-precipitation method or dip-coating method with a C:Fe ratio of 1:30. No differences were discovered in the patterns of Raman spectrum by the Fe catalyst deposition method; however the intensity value was relatively larger when the dip-coating method was used, when compared to the deposition-precipitation method.

As seen in **Figure 9**, a D band appeared around 1332 cm^{-1}, a G band around 1582 cm^{-1}, and a 2D band around 2700 cm^{-1}; and the intensity and $R = (I_D/I_G)$ values of the D band, G band, and 2D band appear in the respective spectra. When CNFs were grown after the deposition of Fe catalyst in the two different methods, the respective R values were 0.64 under deposition-precipitation, while they were 0.73 under the dip-coating method. As such, it is known that CNFs grow better, if and when the Fe catalyst is deposited by dip-coating, and then CNFs are grown. Meanwhile, a 2D band appeared in both samples, which means that not only pure CNFs carbon grow but also multiwalled carbon nanotubes (MWCNTs) partially grow together, when CNFs are grown by both the deposition-precipitation and dip-coating methods [4, 7, 17].

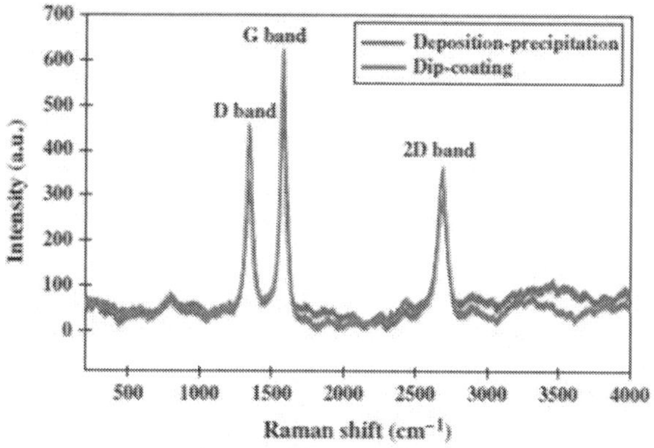

Deposition method of Fe catalyst	*D* band (intensity a.u.)	*G* band (intensity a.u.)	*2D* band (intensity a.u.)	*R* (I_D/I_G)
Deposition-precipitation	323	507	283	0.64
Dip-coating	455	620	364	0.73

Figure 9. The Raman spectra of CNFs grown on C-fiber textiles by the two different deposition methods.

3. Synthesis of CNFs and CNFs/SiO_2 composite on electroplated Co-Ni/C-fiber textiles

3.1. Experimental process

3.1.1. Deposition of Co-Ni catalysts on C-fiber textiles

Using the electrophoretic method, Co-Ni catalysts were deposited on C-fiber textiles. Carbon electrodes and carbon sheets were used as the anode and cathode, respectively, while a mixture of cobalt nitrate and nickel nitrate was used as the electrolyte. To analyze the characteristics of the CNFs according to their cobalt and nickel contents, cobalt nitrate and nickel nitrate, with the weight ratio of 6:4 and 8:2, were used as the electrolyte for the deposition of Co-Ni catalysts by applying 0.04–0.05 A of current for 5 min [4, 7, 17, 18, 22].

3.1.2. Reduction of catalysts and synthesis of CNFs

After placing the C-fiber textiles with deposited catalysts into a furnace, the temperature was increased by 12°C/min up to 700°C in N_2 atmosphere. While maintaining the temperature at 700°C, H_2 gas was flowed into the furnace for 1 hour for the reduction of the catalysts.

Using CVD, CNFs were synthesized to electroplated Co-Ni/C-fiber textiles. Ethylene gas, as the C source, was flowed into the furnace for an hour at 700°C, during which reduction was completed. After the completion of the reaction, the temperature was lowered to room temperature in N_2 atmosphere [4, 7, 17, 18, 22].

3.1.3. Synthesis of the CNFs/SiO_2 composite

For the silica coating, a hydroxyl group was attached to the surface of the CNFs as an anchor group and oxidized for 30 min in 80°C nitric acid. Using the sol-gel process, SiO_2 was applied as a coating through the hydrolysis of tetraethyl orthosilicate (TEOS) on the surface of the reduced CNFs. TEOS was hydrolyzed by dissolving hydrochloric acid and distilled water with ethanol and stirred for 6 hours at room temperature. After soaking the CNFs/C-fiber textiles in this solution and adding ammonia solution for gelation, the solution was stirred for 12 hours and SiO_2 was coated on the surface of the CNFs [4, 7, 17, 18, 22].

3.1.4. Fabrication of coin cell

To investigate the electrochemical characteristics of the CNFs and the CNFs/SiO_2 composite, a coin cell was fabricated. For the working electrode, the CNFs and CNFs/SiO_2 deposited on C-fiber textiles were used without a binder, Li metal was used for the counter electrode, and polyethylene was used for the separator. For the electrolyte, $LiPF_6$ was dissolved in ethylene carbonate (EC) and diethyl carbonate (DEC) (1:1 vol.%) for a mixed solution. The coin cell was fabricated in a glove box filled with Ar gas [4, 7, 17, 18, 22].

3.2. Characterization of the CNFs/SiO_2 composite

3.2.1. Scanning Electron Microscope (SEM)

Figure 10 shows the SEM images of CNFs synthesized according to the Co-Ni ratio. In both samples, CNFs in a curved form with a uniform diameter were synthesized. With a Co-Ni ratio of 8:2, 40 nm CNFs were synthesized, while 30 nm CNFs were synthesized with a ratio of 6:4 [22].

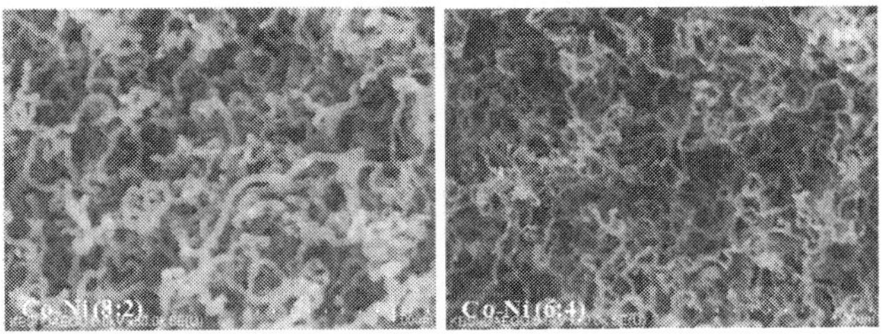

Figure 10. SEM images of CNFs grown on Co-Ni/C-fiber textiles.

3.2.2. X-ray Photoelectron Spectroscopy (XPS)

Figure 11 shows the XPS spectra of CNFs that were synthesized using Co-Ni catalysts. With Co-Ni ratios of 8:2 and 6:4, a C═C (sp^2) bond, C—C (sp^3) bond, —C═O bond, and —COO bond were observed around 285, 286, 288, and 292 eV, respectively. It was found that the catalyst ratio did not affect the binding energy of CNFs as the same C-binding energy was observed with both catalyst ratios.

Figure 12 shows the XPS spectra of the as-synthesized CNFs/SiO_2 composite. Around the binding energy of 103–105 eV for silicon, SiO_2 and SiO were identified. Considering that the intensity of SiO_2 bonding is higher than that of SiO bonding in larger areas, the coated layer on the surface of CNFs consisted mostly of SiO_2 [22].

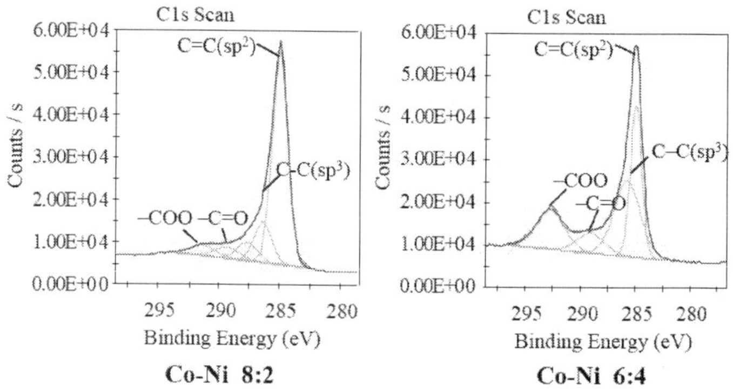

Figure 11. XPS spectra of CNFs grown on Co-Ni/C-fiber textiles.

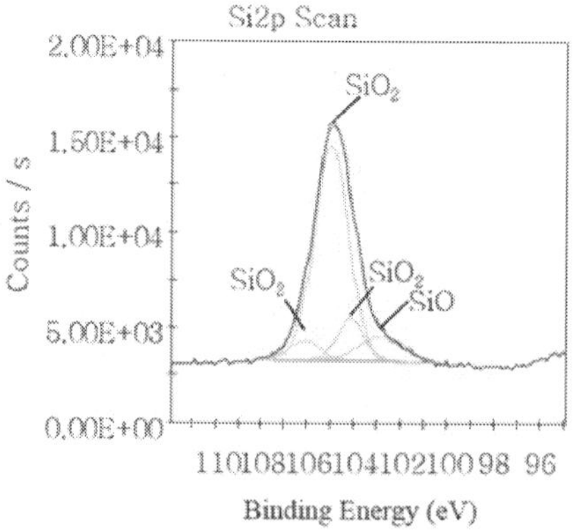

Figure 12. XPS spectra of CNFs/SiO_2 composite.

3.2.3. Raman

Figure 13 shows the Raman spectra of CNFs according to the Co-Ni ratios. Around 1590 cm^{-1}, a G band indicating a C═C (sp^2) bond was observed, while a D band indicating a C—C (sp^3) bond was observed around 1350 cm^{-1}. Considering that the intensity ratio of the D band and B band at both ratios was close to 1, it was determined that CNFs were synthesized with a 1:1 ratio of sp^2 bonds and sp^3 bonds. The intensity of CNFs synthesized at the Co-Ni ratio of 8:2 was higher than that of CNFs synthesized at a 6:4 ratio. Accordingly, this showed that CNFs synthesized at the Co-Ni ratio of 8:2 had higher crystallizability [22].

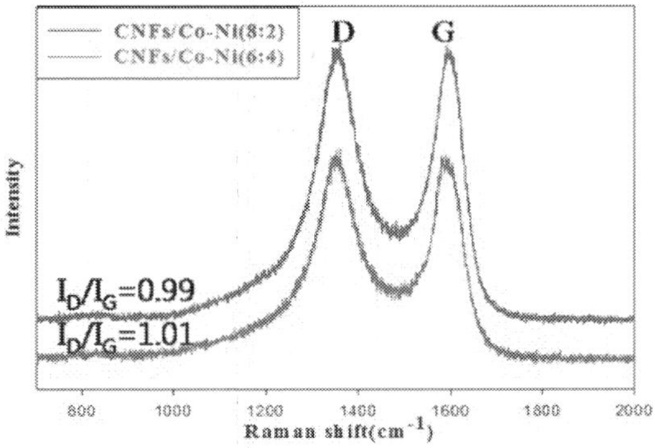

Figure 13. Raman spectra of CNFs grown on Co-Ni/C-fiber textiles.

3.2.4. Transmission electron microscopy (TEM)

To observe the layer of silica coated on the surface of the CNFs, TEM images were examined and are shown in **Figure 14**. For the CNFs coated with silica, CNFs synthesized at the Co-Ni ratio of 8:2 with relatively high crystallizability were used. A layer of silica at a thickness of about 10 nm was coated evenly on the surface of the CNFs in a hollow form with about 8 nm center diameter [22].

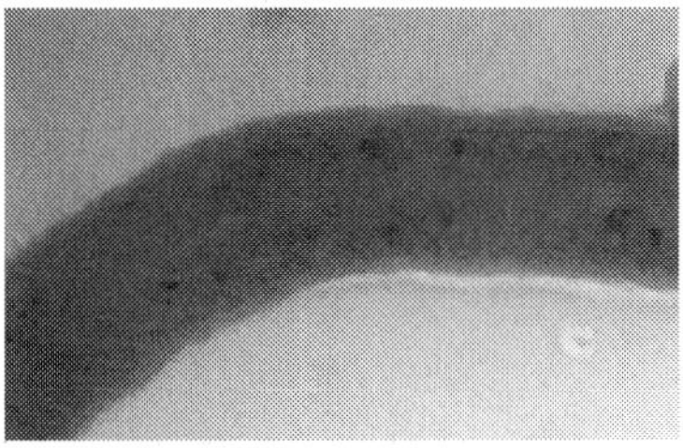

Figure 14. TEM image of SiO_2-coated CNF composite.

3.2.5. Cycle performance

The discharge capacities of C-fiber textiles, CNFs, and CNFs/SiO_2 composite were measured and are shown in **Figure 15**. The respective initial discharge capacities of the CNFs synthesized with the Co-Ni ratios of 8:2 and 6:4 were 258 and 234 mAh g^{-1}, respectively, thereby showing higher results than that of C-fiber textiles with the initial discharge capacity of 190 mAh g^{-1} and a high retention rate of 95%. The CNFs/SiO_2 composite showed the highest initial discharge capacity of 1468 mAh g^{-1} with the retention rate of 47% [22].

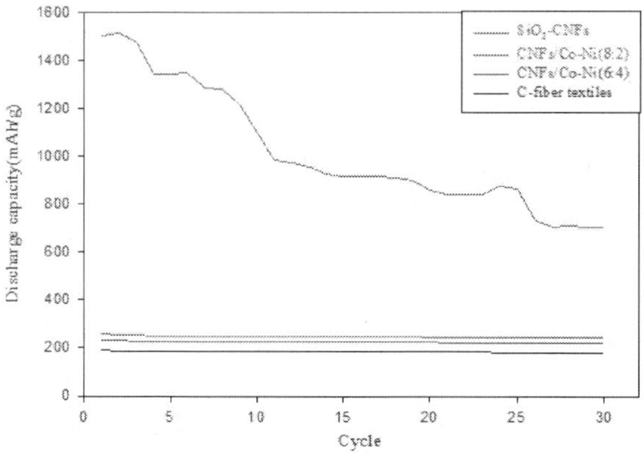

	C-fiber textiles	CNFs/ Co-Ni(8:2)	CNFs/ Co-Ni(6:4)	SiO_2-CNFs
Discharge capacity at 1st cycle	190	258	234	1468
Discharge capacity at 30th cycle	183	244	220	705
Retention rate (%)	96	95	94	47

Figure 15. Discharge capacities and retention rates of C-fiber textiles, CNF/Co-Ni and CNF/SiO_2 composites.

4. Conclusions

Based on the catalysts deposited using the electrophoretic method, CNFs were synthesized on the surface of C-fiber textiles through CVD. Upon hydrolysis of TEOS, silica was coated on the surface of CNFs to synthesize the CNFs/SiO_2 composite. The characteristics of the as-synthesized CNFs and CNFs/SiO_2 composite were analyzed through various spectroscopic methods. Upon fabricating a coin-type half-cell without a binder for the as-synthesized sample, its cycle performance when used as an anode material in Li secondary batteries was investigated. Based on the results, the following conclusions were deduced [22].

1. When the CNFs grew, after deposition of the Fe catalyst using the deposition-precipitation method, the diameters grew up to 40–60 nm and 30–55 nm, when weight ratios of C:Fe were 1:30 and 1:50, respectively.

2. When the CNFs grew, after deposition of Fe catalyst using the dip-coating method, the diameters grew up to 40–60 nm and 25–30 nm, when the weight ratios of C:Fe were 1:10 and 1:30, respectively [4, 7, 17].

3. With Co-Ni ratios of 8:2 and 6:4, CNFs with diameters of 40 and 30 nm were synthesized, respectively [22].

4. The Raman analysis results confirmed that CNFs with a 1:1 ratio of C sp^2 bonds and sp^3 bonds were synthesized when Co-Ni catalysts were used. When the Co-Ni ratio was 8:2, CNFs with higher crystallizability were synthesized compared to that synthesized with a ratio of 6:4 [22].

5. The discharge capacity of the CNFs synthesized on the surface of the C-fiber textiles was 230–260 mAh g^{-1}, about 35% higher than that of pure C-fiber textiles, with a high retention rate of 95% [22].

6. CNFs/SiO_2 composites showed the highest discharge capacity of 1468 mAh g^{-1} with a retention rate of 47% [22].

Acknowledgements

This research was financially supported by the Ministry of Education, Science Technology (MEST) and National Research Foundation of Korea (NRF) through the Human Resource Training Project for Regional Innovation (No. 2015035858).

Author details

Chang-Seop Lee* and Yura Hyun

*Address all correspondence to: surfkm@kmu.ac.kr

Department of Chemistry, Keimyung University, Daegu, South Korea

References

[1] Kenneth B. K. Teo Charanjeet Singh, Manish Chhowalla, Willam I. Milne. Catalytic Synthesis of Carbon Nanotubes and Nanofibers. In: H. S. Nalwa, editor. Encyclope-

dia of Nanoscience and Nanotechnology. 10th ed. USA: American Scientific Publishers; 2003. p. 1–22.

[2] Yoong-Ahm Kim, Takuya Hayashi, Satoru Naokawa, Takashi Yanagisawa, Morinobu Endo. Comparative Study of Herringbone and Stacked-cup Carbon Nanofibers. Carbon. 2005;43:3005–3008. DOI: 10.1016/j.carbon.2005.06.037

[3] Eunyi Jang, Heai-Ku Park, Chang-Seop Lee. Synthesis and Application of Si/Carbon Nanofiber Composites Based on Ni and Mo Catalysts for Anode Material of Lithium Secondary Batteries. Journal of Nanoscience and Nanotechnology. Forthcoming. DOI: 10.1166/jnn.2016.12229

[4] Sang-Won Lee, Chang-Seop Lee. Electrophoretic Deposition of Iron Catalyst on C-Fiber Textiles for the Growth of Carbon Nanofibers. Journal of Nanoscience and Nanotechnology. 2014;14:8619–8625. DOI: 10.1166/jnn.2014.9960

[5] Yura Hyun, Eun-Sil Park, Karina Mees, Ho-Seon Park, Monika Willert-Porada, Chang-Seop Lee. Synthesis and Characterization of Carbon Nanofibers on Transition Metal Catalysts by Chemical Vapor Deposition. Journal of Nanoscience and Nanotechnology. 2015;15:7293–7304. DOI: 10.1166/jnn.2015.10588

[6] Eun-Sil Park, Jong-Won Kim, Chang-Seop Lee. Synthesis and Characterization of Carbon nanofibers on Co and Cu Catalysts by Chemical Vapor Deposition. Bulletin of the Korean Chemical Society. 2014;35(6):1687–1691. DOI: 10.5012/bkcs.2014.35.6.1687

[7] Sang-Won Lee, Chang-Seop Lee. Growth and Characterization of Carbon Nanofibers on Fe/C-Fiber Textiles Coated by Deposition–Precipitation and Dip-Coating. Journal of Nanoscience and Nanotechnology. 2015;15:7317–7326. DOI: 10.1166/jnn.2015.10585

[8] Eunyi Jang, Heai-Ku Park, Jong-Ha Choi, Chang-Seop Lee. Synthesis and Characterization of Carbon Nanofibers Grown on Ni and Mo Catalysts by Chemical Vapor Deposition. Bulletin of the Korean Chemical Society. 2015;36(5):1452–1459. DOI: 10.1002/bkcs.10285

[9] Yura Hyun, Jin-Yeong Choi, Heai-Ku Park, Jae Young Bae, Chang-Seop Lee. Synthesis and Electrochemical performance of Mesoporous SiO_2-Carbon nanofibers composite as anode materials for lithium secondary batteries. Materials Research Bulletin. Forthcoming. DOI:10.1016/j.materresbull.2016.03.006

[10] Xiaojiao Li, Chuncheng Hao, Qingquan Lei. Growth of Carbon Nanofibers Catalyzed by Silica-coated Copper Nanoparticles. Materials Research Bulletin. 2012;47:352–355. DOI: 10.1016/j.materresbull.2011.11.013

[11] Kyung Ho Park, Soonil Lee, Ken Ha Koh. Growth and High Current Field Emission of Carbon Nanofiber Films with Electroplated Ni Catalyst. Diamond and Related Materials. 2005;14(11–12):2094–2098. DOI: 10.1016/j.diamond.2005.06.013

[12] Martijn K. van der Lee, A. Jos van Dillen, John W. Geus, Krijn P. de Jong, Johannes H. Bitter. Catalytic Growth of Macroscopic Carbon Nanofiber Bodies with High Bulk

Density and High Mechanical Strength. Carbon 2006;44(4):629–637. DOI: 10.1016/j.carbon.2005.09.031

[13] Young Joon Yoon, Hong Koo Baik. Catalytic Growth Mechanism of Carbon Nanofibers Through Chemical Vapor Deposition. Diamond and Related Materials. 2001;10(3–7):1214–1217. DOI: 10.1016/S0925-9635(00)00585-9

[14] Jae-Seok Lim, Seong-Young Lee, Sei-Min Park, Myung-Soo Kim. Preparation of Carbon Nanofibers by Catalytic CVD and Their Purification. Carbon Letters. 2005;6(1):31–40.

[15] Taeyun Kim, Karina Mees, Ho-Seon Park, Monika Willert-Porada, Chang-Seop Lee. Growth of Carbon Nanofibers Using Resol-type Phenolic Resin and Cobalt(II) Catalyst. Journal of Nanoscience and Nanotechnology. 2013;13(11):7337–7348. DOI: 10.1166/jnn.2013.7852

[16] Yura Hyun, Haesik Kim, Chang-Seop Lee. Synthesis of Carbon Nanofibers on Iron and Copper Catalysts by Chemical Vapor Deposition. Advanced Materials Research. 2013;750–752:265–275. DOI: 10.4028/www.scientific.net/AMR.750-752.265

[17] Sang-Won Lee, Karina Mees, Ho-Seon Park, Monika Willert-Porada, Chang-Seop Lee. Synthesis of Carbon Nanofibers on C-Fiber Textiles by Thermal CVD Using Fe Catalyst. Advanced Materials Research. 2013;750–752:280–292. DOI: 10.4028/www.scientific.net/AMR.750-752.280

[18] Ki-Mok Nam, Heai-Ku Park, Chang-Seop Lee. Synthesis and Electrochemical Properties of Carbon Nanofibers and SiO_2/Carbon Nanofiber Composite on Ni–Cu/C-Fiber Textiles. Journal of Nanoscience and Nanotechnology. 2015;15:8989–8995. DOI: 10.1166/jnn.2015.11555

[19] Eun-Sil Park, Heai-Ku Park, Ho-Seon Park, Chang-Seop Lee. Synthesis and Electrochemical Properties of CNFs–Si Composites as an Anode Material for Li Secondary Batteries. Journal of Nanoscience and Nanotechnology. 2015;15:8961–8970. DOI: 10.1166/jnn.2015.11554

[20] Yura Hyun, Heai-Ku Park, Ho-Seon Park, Chang-Seop Lee. Characteristics and Electrochemical Performance of Si-Carbon Nanofibers Composite as Anode Material for Binder-Free Lithium Secondary Batteries. Journal of Nanoscience and Nanotechnology. 2015;15:8951–8960. DOI: 10.1166/jnn.2015.11553

[21] Eun-Sil Park, Jong-Ha Choi, Chang-Seop Lee. Synthesis and Characterization of Vapor-grown Si/CNF and Si/PC/CNF Composites Based on Co–Cu Catalysts. Bulletin of the Korean Chemical Society. 2015;36(5):1366–1372. DOI: 10.1002/bkcs.10262

[22] Kun-Ho Jang, Sang-Hoon Lee, Yujin Han, Seong-Ho Yoon, Chang-Seop Lee. Synthesis and Characteristics of Silica-coated Carbon Nanofibers on Electroplated Co–Ni/C-fiber Textiles. Journal of Nanoscience and Nanotechnology. Forthcoming.

Section 5

State-of-the-Art Nanomaterials

Chapter 9

Electrospun Metal Oxide Nanofibers and Their Energy Applications

Yingjie Liao, Takeshi Fukuda and Suxiao Wang

Additional information is available at the end of the chapter

http://dx.doi.org/10.5772/63414

Abstract

Metal oxide nanofibers have attracted considerable research interest for processing both one-dimensional nanometer scale morphology and unique chemical and electrical properties. A variety of their practical applications in light-emitting diodes, liquid crystal displays, solar cells, and gas sensors have been demonstrated. Electrospinning provides a rapid and facile way to fabricate nanofibers with diameter several orders of magnitude smaller than that produced by conventional spinning methods. In this chapter, we discuss the fabrication of ultrathin metal oxide nanofibers by the electrospinning technique. Priority is given to zinc oxide nanofibers. Major parameters affecting the morphology and diameter of the nanofibers are investigated systematically. The effect of calcination condition on chemical composition and crystallization of the electrospun nanofibers is also addressed. In addition, we show the advantages and problems when applying electrospun nanofibers to solar cells.

Keywords: electrospinning, metal oxide nanofibers, solar cells, energy, zinc oxides

1. Introduction

Nanofibers have been widely used in many areas, including catalysis [1–3], tissue engineering [4–7], filtration [8–10], biomedical [11–13], sensors [14–16], energy [17–19], and magnetic [20, 21]. The most common ceramic materials of nanofibers are zinc oxide (ZnO), titanium dioxide, silicon dioxide, zirconium dioxide, aluminium oxide, and lithium titanate [22–24]. Nanofibers can be synthesized by various methods, such as melting processing, antisolvent-induced polymer precipitation, electrospinning, island in the sea, or direct drawing from a solution [25–27]. Electrospinning is one of the mostly studied techniques due to its simplicity of experimental setup and relatively high production rate. As shown in **Figure 1**, the number of published papers on electrospun nanofibers every year increases quickly in the past 10 years.

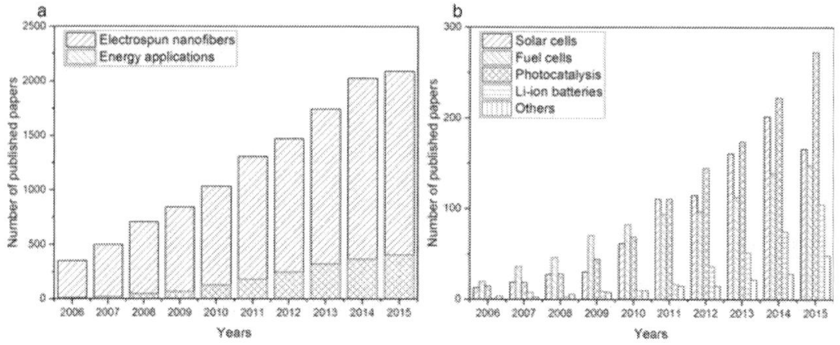

Figure 1. Comparison of the number of research publications (a) on electrospun nanofibers and their energy applications part and (b) on major energy applications of nanofibers in the past 10 years. The data are obtained from Web of Science Core Collection.

As the demand for energy is growing rapidly in recent years, energy conversion and energy storage become hot research topics. Nanofibers and other one-dimensional structures, such as nanowires, nanorods, nanobelts, and nanotubes, have been investigated intensively for their unique properties in energy applications [28–31]. As indicated in **Figure 1a**, about 20% of publications on electrospun nanofibers are for energy applications in 2015. Nanofibers, especially metal oxide nanofibers, possess unique optical and electrical characteristics and have been used in various energy devises [2, 32]. The number of published research articles on the energy application of nanofibers shows exponential increment in the past 10 years. Most of them focus on solar cells, fuel cells, lithium ion batteries, and photocatalysis applications. Other applications, e.g., super capacitors and hydrogen generation, are also attracted a lot of research interest.

In this chapter, we introduce the electrospinning technique briefly and discuss the processing parameters for controlling the diameter and morphology of electrospun metal oxide nanofibers as exemplified by ZnO nanofibers. Then, we show applications of nanofibers in energy conversion (e.g., energy devices and fuel cells) and energy storage (e.g., lithium ion batteries) and figure out the challenges for these applications.

2. Electrospinning of ultrathin metal oxide nanofibers

2.1. Electrospinning technique

The electrospinning technique is considered as one of the most simple and effective ways to fabricate nanofibers. **Figure 2** shows a schematic diagram of an electrospinning setup. A basic electrospinning setup usually consists of three parts: a high-voltage power supply, a spinneret, and a collector. To facilitate monitoring the state of the electrospinning, a laser and a CCD camera are sometimes added into the setup. The inset of **Figure 2** is obtained from a CCD camera and illustrates a typical image of a spinning jet during the electrospinning. A common high-voltage power supply for electrospinning is a direct current power in the range of several kV to about 20 kV. The spinneret is usually connected with a syringe to hold more solution and control the flow rate. The inner diameter of the spinneret can be smaller and more individual for glass capillaries compared to metallic needles. The collector could be an aluminium foil or other substrates with conductive surface, which depends on practical application.

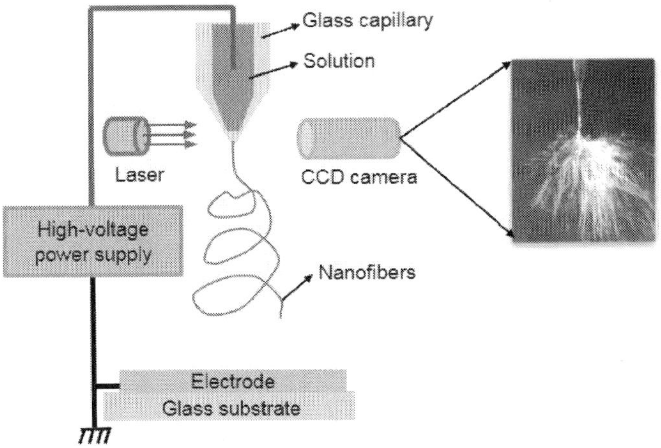

Figure 2. A schematic diagram of the electrospinning apparatus. Inset shows the CCD image of electrospinning process near the tip of the glass capillary.

The electrospinning setup is rather simple and can be built easily, however, the mechanism of electrospinning is very complicated [18, 27]. When a high voltage is applied to the liquid inside the spinneret, an electrical filed is set up between the spinneret and the collector. The liquid (usually a polymer solution) reaching the spinneret tip forms a Taylor cone, which emits a liquid jet through its apex. This electrified liquid jet undergoes a stretching and whipping process, which leads to the formation of a long and thin thread. As the liquid jet is continuously elongated and the solvent is evaporated, its diameter can be greatly reduced from hundreds of micrometres to as small as tens of nanometres. Attracted by the grounded collector placed under the spinneret, randomly orientated fibres are deposited [24].

Nanofibers with novel structures can be archived by modifying the basic electrospinning setup in a number of ways. Modifications are usually in the spinneret or the collector. It has been demonstrated that a spinneret containing two needles could be used to electrospun composite nanofibers. In another approach, a spinneret consisting of two coaxial capillaries has been used to fabricate fibres with core/sheath or hollow structures. In addition, the orientation of electrospun nanofibers can be controlled by changing the electrode of the collector. One also can get relatively uniform mat by replacing the planar collector with a rotating drum.

2.2. Controlling the diameter and morphology of electrospun nanofibers

The diameter and morphology of electrospun nanofibers are two key factors, which should be controlled for practical applications. They are dependent on a lot of processing parameters relevant to the solution and setup conditions. The intrinsic properties of the solution play important role in determining the diameter and morphology of resultant nanofibers. One can get target nanofibers by changing the type of the polymer, the conformation of polymer chain, or by adjusting concentration, viscosity, elasticity, electrical conductivity, and the polarity and surface tension of the solution. The conditions of electrospinning setup, e.g., the flow rate for the solution, the applied voltage, the distance between the spinneret and the collector, could also affect tremendously the diameter and the morphology of electrospun nanofibers. In addition to these parameters, variables of evaporation, such as temperature and humidity, may also influence the two key factors.

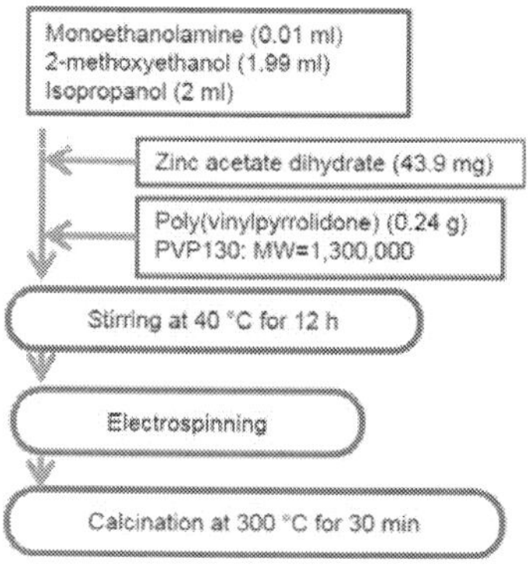

Figure 3. A typical procedure for preparing the ZnO nanofibers by electrospinning.

Here, we take the electrospinning of ZnO nanofibers as an example to explain how to control the diameter and morphology of electrospun nanofibers. **Figure 3** shows a typical procedure for the electrospinning of ZnO nanofibers. In this progress, 43.9 mg of $Zn(CH_3COO)_2{\bullet}2H_2O$ (zinc acetate dihydrate) was first dissolved in a monoethanolamine (MEA)-2-methoxyethanol solution at room temperature. The molar ratio of MEA to zinc acetate dihydrate was kept at 1.0, and the concentration of zinc acetate was 0.1 mol/L. The resultant mixture was stirred at 40°C for 12 hours to obtain a transparent and homogeneous solution. Then polyvinylpyrrolidone (PVP) was added into the ZnO sol-gel solution to increase the viscosity of the solution. The solution was finally loaded into a glass capillary with a 100 μm inner diameter at the blunt tip. A stable high voltage of 6.0 kV was generated by a high-voltage power supply and was applied to the solution through a copper wire in the glass capillary. In addition, an indium tin oxide (ITO)–coated glass substrate (25 mm × 25 mm) was placed perpendicular to the axis of the capillary at a distance of 10 cm from its tip as a collector for solar cell applications (see next section). The collector was connected to the ground along with the high-voltage power supply. The effect of three processing parameters relevant to the properties of the precursor solution, molecular weight of PVP, the concentrations of zinc acetate and PVP in the precursor solution, on the diameter and morphology of electrospun ZnO nanofibers is investigated systematically.

2.2.1. *Effect of the concentration of precursor solution*

SEM images in **Figure 4** show the progression of ZnO-PVP composite from grains to nanofibers when the concentration of PVP in the precursor solution increases from 0.02 to 0.06 g/mL. It is rather clear that only ZnO-PVP composite grains with various sizes are formed with 0.02 g/ml PVP regardless of the concentration of zinc acetate. This implies that the electrospun product is grains instead of nanofibers when the PVP in the solution could not provide enough viscosity for electrospinning.

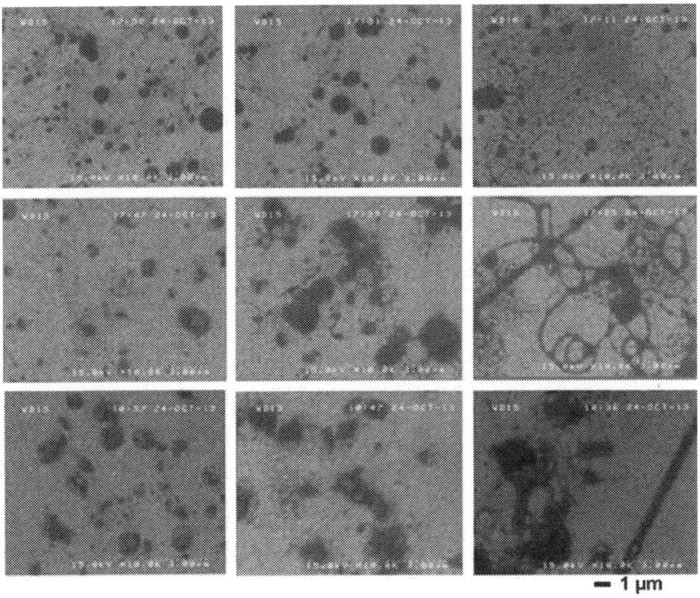

Figure 4. SEM images of the ZnO-PVP composite structure electrospun from a mixture of ZnO sol-gel and PVP solution. Concentrations of zinc acetate are 0.1 M (top row), 0.4 M (middle row), and 0.75 M (bottom row); those of the PVP solution are 0.02, 0.04, and 0.06 g/mL from the left to the right column, respectively [33].

As the concentration of PVP is increased to 0.04 g/mL, a few ZnO-PVP nanofibers appear among ZnO-PVP grains, see the middle column of **Figure 4**. When the concentration of PVP reaches 0.06 g/mL, ZnO-PVP nanofibers become predominant (right column of **Figure 4**). It is worth noting that similar progression from grains to nanofibers is taken place in all solutions (0.1 M, 0.4 M and 0.75 M zinc acetate). So, we can conclude that it is not the concentration of zinc acetate but the concentration of PVP, which determines the formation of ZnO-PVP nanofibers.

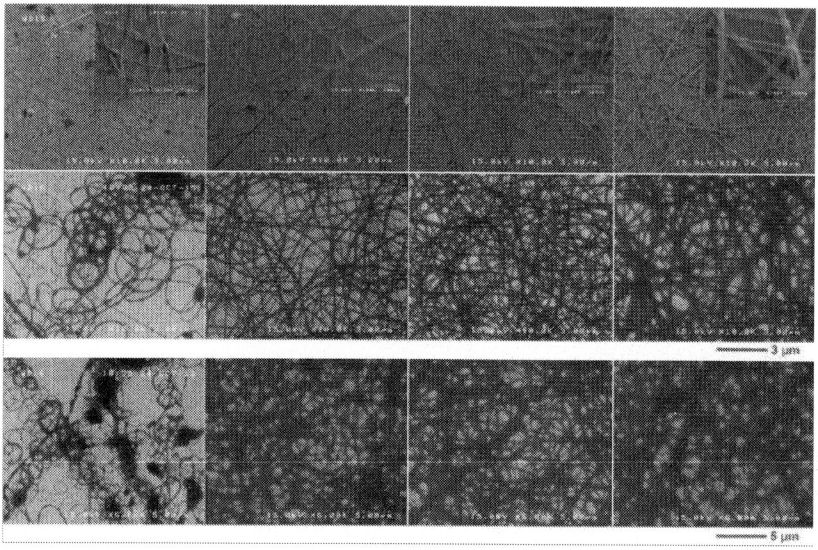

Figure 5. SEM images of the ZnO-PVP composite nanofibers electrospun from a mixture of ZnO sol–gel and PVP solution. Concentrations of zinc acetate are 0.1 M (top row), 0.4 M (middle row), and 0.75 M (bottom row); those of the PVP solution are 0.06, 0.08, 0.12, and 0.14 g/mL from the left to the right column, respectively. High-magnification SEM images are shown as insets [33].

SEM images in **Figure 5** show the effect of the concentration of precursor solution on the diameter of the ZnO-PVP composite nanofibers. One can see that the diameter of electrospun nanofibers increases steadily with the concentration of PVP for all three groups of samples. And the beads present in the top row images (0.1 M zinc acetate) become less prominent with the growth of the nanofibers, which can also be attributed to the increment of the viscosity of the precursor solution induced by more PVP [27]. These results suggest that the concentration of PVP in the precursor solution plays a significant role in determining not only the size of the nanofibers but also the absence of the beads. As for the concentration of zinc acetate, we find that it could also influence the diameter of resultant nanofibers when comparing the three groups of samples in **Figure 5**. Relatively high molar concentration of zinc acetate induces the formation of thicker ZnO-PVP composite nanofibers. Moreover, the nanofibers synthesized with 0.1 M zinc acetate are more uniform than those in the other two groups. In general, the diameter of ZnO-PVP nanofibers can be controlled through changing the concentration of either PVP or zinc acetate in the precursor solution. But thin ZnO-PVP nanofibers are produced commonly with some beads at low concentration of the precursor solution.

In order to analyze quantitatively and compare the effect of the concentration of PVP and zinc acetate on the size of the resultant nanofibers, we measured the diameter of the nanofibers from their high-resolution SEM images and plotted the mean of 50 measurements with corresponding standard error for each sample (**Figure 6**).

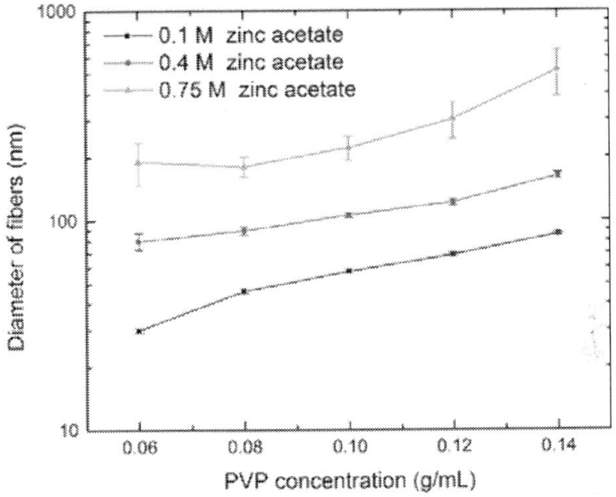

Figure 6. Statistics for the diameter of the ZnO-PVP nanofibers [33].

For the ZnO nanofibers synthesized with 0.1 M zinc acetate, their diameter increase almost linearly with the concentration of PVP. The finest ZnO nanofiber is 29.9 ± 0.8 nm in diameter. In the case of 0.4 M zinc acetate, the diameter of electrospun ZnO nanofibers increased superlinearly from 79.9 ± 7.1 to 162.0 ± 5.5 nm as the PVP concentration increased from 0.06 to 0.14 g/mL. Comparing the fibres synthesized with given PVP concentration, we found that their diameter increases considerably with the molar concentration of zinc acetate. We also noticed that the standard error of the mean diameter for the fibres synthesized with 0.4 and 0.75 M zinc acetate, especially the latter, is larger than the case of 0.1 M zinc acetate, which implies that the concentrated ZnO sol-gel solution disturbed the balance electrospinning set up by the PVP component. In general, one can get ZnO nanofibers with large diameter by increasing the concentration of zinc acetate or PVP in the precursor solution. And the former could contribute to greater nonuniformity in the distribution of the diameter than the latter.

2.2.2. *Effect of the molar mass of polymer and applied voltage*

As indicated by the Mark-Houwink equation, the intrinsic viscosity of a polymer is related to its molecular weight. That means one can adjust the viscosity of a polymer solution by changing the molar mass of the polymer. The parameters of electrospinning setup could also influence the diameter and morphology of electrospun nanofibers. Here, the molar mass of PVP and applied voltage are investigated together.

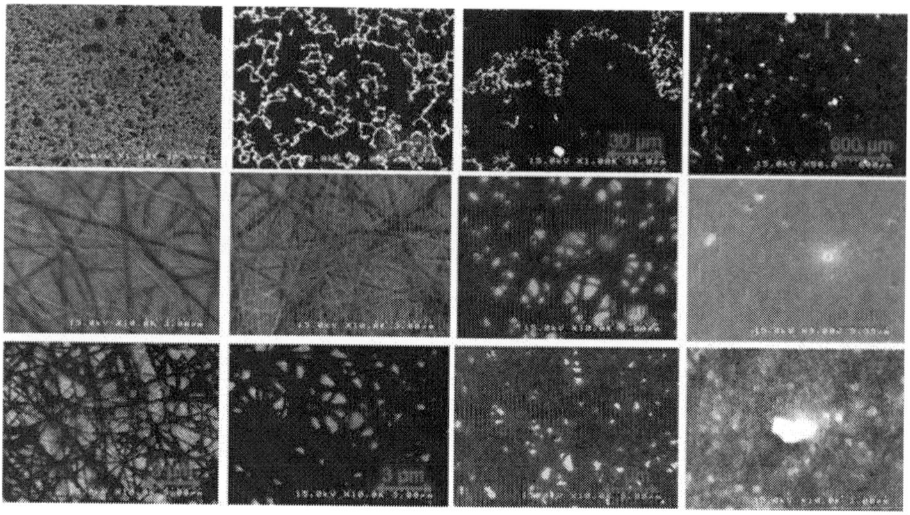

Figure 7. SEM images of ZnO-PVP nanofibers electrospun with PVP of different molar mass (top: 40,000; middle: 360,000; bottom: 1,300,000). And the high voltage applied to the spinneret is 6, 7, and 8 kV from left to right.

One can see from the first row in **Figure 7**, with PVP 40,000, the electrospun product is varicose fibres with relative large size and short length. As the applied voltage increases from 6 to 8 kV, the size of fibres become larger, but the amount of fibres is fewer. For PVP 360,000 and PVP 1,300,000, ZnO-PVP nanofibers are observed. These nanofibers get denser with increment of the applied voltage. On the whole, the role of the molar mass of PVP in determining the diameter of ZnO-PVP nanofibers is similar to that of the concentration of precursor solution. At the same time, turning the concentration is much easier than changing the molar mass for practical application. The applied voltage could also affect the size of electrospun nanofibers, but not as direct as other parameters discussed above.

2.3. Crystallization of metal oxide nanofibers

Electrospinning of composite nanofibers is the first step for the preparation of metal oxide nanofibers. In order to improve the purity and other characteristics of the composite nanofibers (e.g., transmittance, mobility, and crystallinity) for energy application, calcination will be conducted after electrospinning. For example, the as-spun ZnO-PVP nanofibers described above could be subsequently converted into ZnO nanofibers with pure phase through calcination. In this section, we discuss the calcination conditions for the crystallization of metal oxide nanofibers.

http://dx.doi.org/10.5772/63414

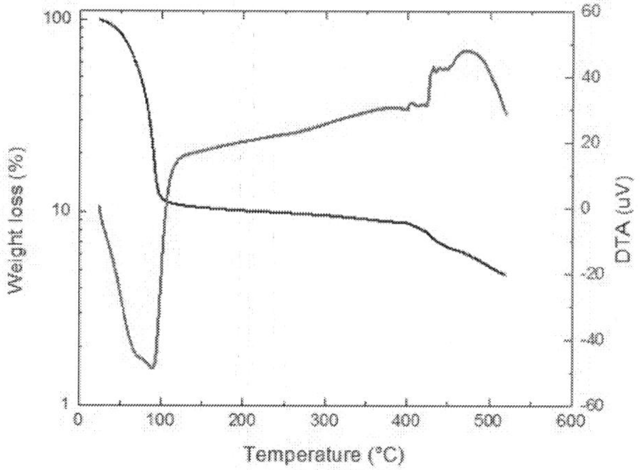

Figure 8. Thermogravimetric analysis curves of as-spun ZnO-PVP nanofibers.

To determine the decomposition temperature of the precursor and the calcination temperature for the ZnO-PVP nanofibers, thermogravimetric analysis for the as-spun ZnO-PVP nanofibers was performed and shown in **Figure 8**. For the thermogravimetric analysis curve, the residue weight of the as-spun ZnO-PVP nanofibers is only 4.73%. Most organic in the as-spun ZnO-PVP nanofibers comes from PVP, and the CH_3COOH group of zinc acetate was removed at a temperature below 480°C. A strong endothermic peak appeared at around 100°C in the differential thermal analysis (DTA) curve, which corresponds to the loss of absorbed water. The exothermic peak at around 400, 430, and 463°C in the DTA curve may have resulted from decomposition on the side and main chains of PVP. The thermogravimetric analysis indicates that a calcination temperature of the as-spun ZnO-PVP nanofibers may existed above 480°C.

Transmission electron microscopy (TEM) images in **Figure 9** show the microscopic structure of ZnO nanofibers after calcination under different conditions. The diameter of as-spun ZnO-PVP nanofibers is around 120 nm before calcination. After calcination at 300°C for 10 minutes, the diameter of the nanofibers does not change a lot, see **Figure 9a**. And the shape of the nanofibers is rather uniform. Even from the magnified image of **Figure 9a** (see **Figure 9b**), it is difficult to identify ZnO grains, which suggests that the ZnO did not crystallize sufficiently. From the thermogravimetric analysis above, one can know that the PVP component in the as-spun ZnO nanofibers did not decompose at 300°C. This means that the as-spun ZnO-PVP nanofibers need higher calcination temperature and longer calcination duration to remove the PVP content and crystallize as-spun ZnO-PVP nanofibers. **Figure 9c** shows the TEM image of the as-spun ZnO-PVP nanofibers calcined at 500°C for 2 hours. The diameter of the nanofiber shrinks down to about 50 nm. It is clear that the nanofiber is composed of many single isolated ZnO grains. The lattice images of ZnO in **Figure 9d** suggest that ZnO crystallized under this calcination condition, which is also evidenced by the X-ray diffraction (XRD) analysis given below. The growth direction of the crystalline ZnO is indicated by a red arrow in **Figure 9d**. By comparing the two samples of ZnO-PVP nanofiber calcined under different conditions, one can learn that as-spun ZnO-PVP nanofibers could be converted into crystallized ZnO nanofibers by calcination at 500°C for 2 hours in air. Calcination at 300°C for 10 minutes is insufficient for the decomposition of PVP and crystallization of ZnO-PVP composite.

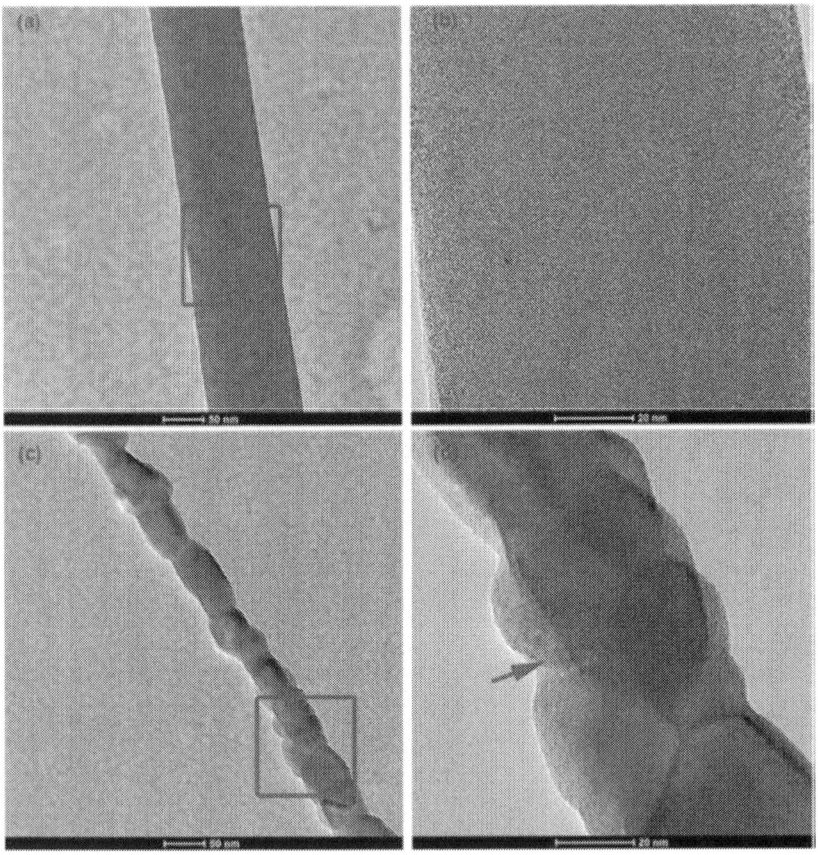

Figure 9. TEM images of the fibres electrospun from the solution of 0.1 M zinc acetate and 0.12 g/mL PVP. After calcination (a, b) at 300°C for 10 min and (c, d) at 500°C for 2 hours [33].

Figure 10 shows the XRD patterns of ZnO-PVP composite nanofibers after calcination at 300°C for 10 minutes and after calcination at 500°C for 2 hours, respectively. The XRD pattern for the sample calcined at 300°C for 10 minutes indicates amorphous ZnO in nature. Characteristic peaks for ZnO are rather weak and obscure, which indicates that only few portions of crystalline ZnO are present under this calcination condition. Strong diffraction peaks in the pattern come from the substrate used for XRD measurement. As for the sample with calcination at 500 °C for 2 hours, five diffraction peaks at 31.76°, 34.34°, 36.20°, 56.50°, and 62.84° appear corresponding to (100), (002), (101), (110), and (103) of the wurtzite crystal structure, respectively. These diffraction peaks can be indexed to a wurtzite hexagonal phase of ZnO. And the phase of the ZnO nanofibers obtained after calcination at 500°C for 2 hours is rather pure, because no characteristic peaks for other impurities were found in the XRD pattern. So, it can be conclude that the calcination condition plays an important role in crystallizing the as-spun ZnO-PVP nanofibers and removing residual PVP component from the nanofibers.

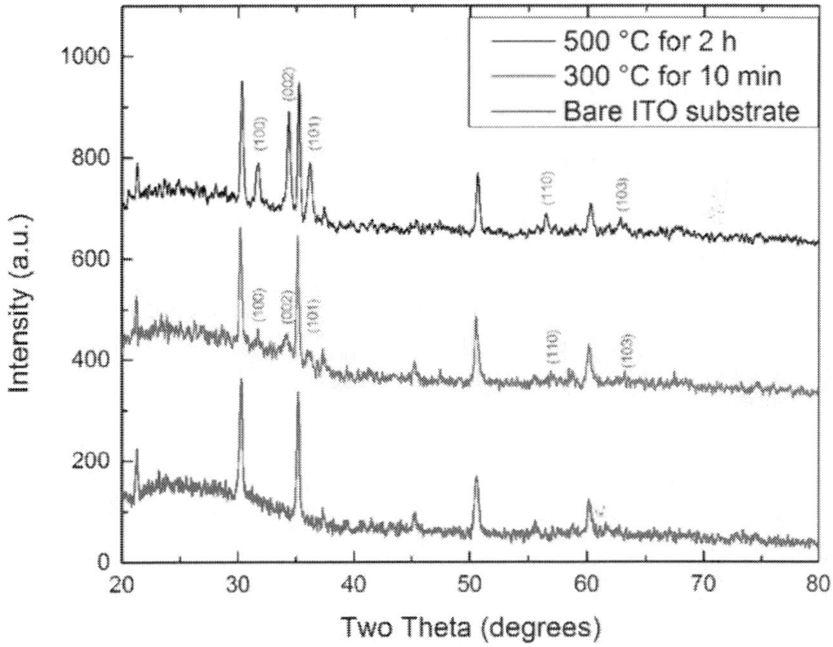

Figure 10. XRD pattern of ZnO nanofibers after calcination under different conditions. The ZnO nanofibers were synthesized using the precursor solution containing 0.1 M zinc acetate and 0.14 g/mL PVP [33].

3. Applications of electrospun nanofibers in the fields of energy

Electrospun nanofibers possess large surface area, high chemical reactivity, and relatively low density, which could contribute a higher absorption of light and dye than the bulk counterparts. These unique optical and electronic properties enable them to be widely applied to a variety of energy harvesting and storage devices, such as solar cells, photocatalysts, fuel cells, and lithium-ion batteries.

3.1. Solar cell applications

Electrospun nanofibers are commonly made from metal oxides, composites, or surface modification. Metal oxide nanofibers, e.g., ZnO nanofibers and Ti_2O nanofibers, are some of the mostly used nanofibers in solar cell applications. One can archive high power conversion efficiency by fast charge transfer and efficient charge separation introduced by nanofibers. Many polymers, such as PVP, polymethylmethacrylate, and polystyrene, are added into the solution to increase the viscosity for the electrospinning of metal oxide nanofibers. Among these polymers, PVP is the most popular one for its high solubility in various solvents and possibility in mass production [34]. It can used to substitute the liquid state electrolyte for solid or semi-solid electrolyte for dye-sensitized solar cell devices. Hybrid solar cells and dye-sensitized solar cells are two major applications of nanofibers in solar cells, which will be introduced as following.

3.1.1. Hybrid solar cells

Hybrid solar cells have attracted considerable research interest as a promising type of solar cells in which the merits of organic and inorganic materials combine. Commonly used combination of organic and inorganic material in hybrid solar cells is polymers and metal oxides. Both of them play important role in improving the performance of hybrid solar cells [35]. The polymers, e.g., poly(3-hexylthiophene-2,5-diyl) (P3HT) and phenyl-C61-butyric acid methyl ester (PCBM), could constitute a donor-acceptor system just like they work in polymer solar cells. The inorganic materials (e.g., ZnO and Ti_2O) could help improve electron transport for the donor-acceptor system, which is essential for archiving high power conversion efficiency. It has been reported that nanofibers of polymer-metal oxide composites could be fabricated by the electrospinning technique. One-dimensional nanofibrous structure provides much larger interface than the conventional planar structure of polymers and metal oxides. Researcher recently found that the usage of metal oxide nanofibers in hybrid solar cells could not only increase the interfacial area but also contribute to the improvement of electron mobility and the dissociation of excitons. The external quantum efficiency could be improved tremendously at the absorption peak of P3HT due to the enhanced dissociation of photo-generated excitons when an electrospun ZnO nanofiber layer is inserted into the P3HT:PCBM blend. Hybrid solar cells using ZnO nanofibers and a P3HT:PCBM donor-acceptor system could yield a power conversion efficiency over 2% [36].

Our group has worked on hybrid solar cells using ZnO nanofibers as electron transport materials. **Figure 11** presents a scheme for the structure of devices used. ZnO nanofibers were fabricated as described in previous section. A blend of P3HT:PCBM is added into the ZnO nanofiber mat. To avoid direct contact between the P3HT:PCBM blend layer and the indium tin oxide (ITO) electrode, a thin ZnO layer was spin coated on the ITO-coated glass substrate before electrospinning of ZnO nanofibers. Here, MoO_x was used as the anode interfacial layer instead of commonly used mixture of poly(3,4-ethylenedioxythiophene) and polystyrene sulfonate, because it has higher charge extraction efficiency.

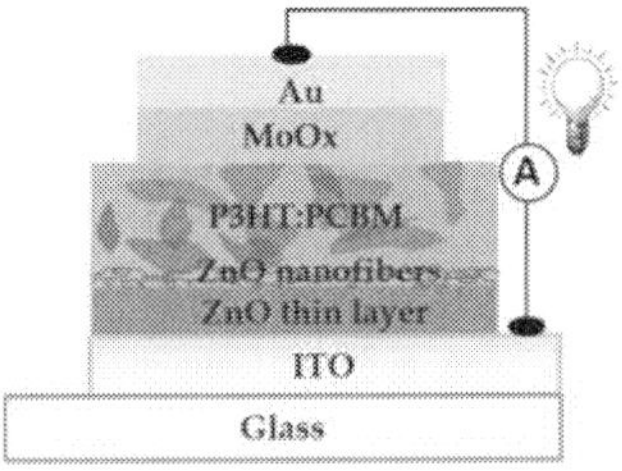

Figure 11. Device structure of hybrid solar cells using ZnO nanofibers as an electron-transporting layer.

To crystallize the electrospun ZnO nanofibers, calcinations at 300 and 500°C were performed. The nanoscale surface morphologies of calcined ZnO nanofibers were investigated by atomic force microscopy (AFM). As shown in **Figure 12**, ZnO nanofibers can be identified from the surface of ITO substrate after calcination. As the calcination temperature increases from 300 to 500°C, more particles can be observed. As confirmed by TEM and XRD (see **Figures 9** and

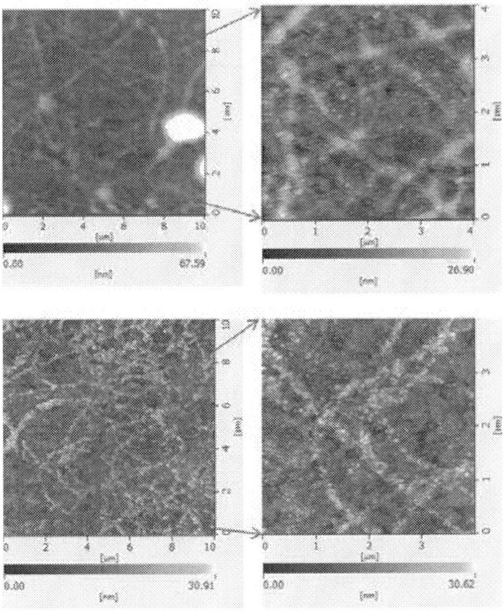

Figure 12. AFM topographies of ZnO nanofibers after calcination at 300°C for 10 min (top) and at 500°C for 2 hours (bottom).

10), these particles are crystalline ZnO. From the magnified AFM images in **Figure 12**, one can also find that the PVP component on the surface of the nanofibers was partially removed after calcination at 500°C. The surface of ZnO nanofibers, however, could still be compositionally rich in PVP even after calcination. The residual PVP may create a contact barrier between the ZnO nanofibers and the P3HT:PCBM blend resulting in a low charge collection efficiency.

Previous paper has shown that the PVP on the surface of colloidal surface could be removed by UV-ozone treatment [37, 38]. To ensure a good contact between ZnO nanofibers and the active layer, we performed UV-ozone treatment on ZnO nanofibers to remove residual PVP from the nanofibers. The atomic concentration of zinc, carbon, and oxygen in the UV-ozone–treated ZnO nanofibers based on the Zn *2p*, C *1s*, and O *1s* XPS spectra is summarized in **Figure 13**. The atomic concentration of zinc and oxygen present in the UV-ozone–treated ZnO nanofibers both increase as the time of treatment. The atomic concentration of oxygen shows relatively larger increment than zinc because there is a competition between increment and decrement in the oxygen atomic concentration by UV-ozone treatment and removal of PVP, respectively. On the contrary, the atomic concentration of carbon from the PVP in the ZnO nanofibers reduced by the UV-ozone treatment, especially from 5 to 7.5 min.

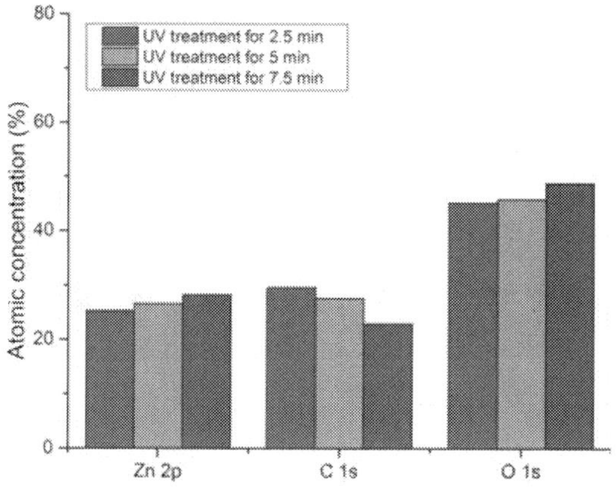

Figure 13. Atomic concentrations of zinc, carbon, and oxygen based on the corresponding XPS spectra. The ZnO nanofibers were measured after UV-ozone treatment for 2.5, 5, and 7.5 min.

The performance result for the hybrid solar cells with UV-ozone–treated ZnO nanofibers is shown in **Figure 14**. The J-V characteristics for the hybrid solar cells were measured under AM 1.5G solar illumination at 100 mW cm^{-2}. The ZnO nanofibers were UV-ozone treated for 2.5, 5, and 7.5 min, resulting in a variation in the J_{sc}. Although the J_{sc} decreases from 2.5 to 5 min, it increases a lot at 7.5 min. As for the UV-ozone treatment for a longer time, it may reduce the electron extraction efficiency due to excess oxygen on the surface of the ZnO nanofibers. It is worth noting that UV-ozone treatment could somewhat contribute to an improvement of charge collection but the power conversion efficiency of the hybrid solar cells with UV-ozone–

treated ZnO nanofibers is lower than the reference device. Various chemical surface treatments are investigated for the application of ZnO nanofibers in hybrid solar cells.

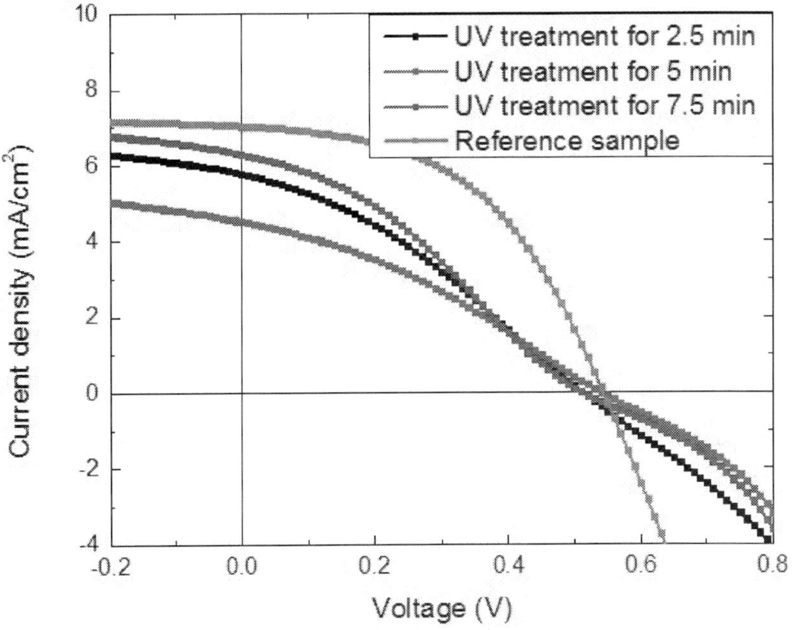

Figure 14. J-V characteristics for hybrid solar cells with UV-ozone–treated ZnO nanofibers for various treatment times (2.5, 5, and 7.5 min).

PVP can dissolve in water and a lot of polar solvents. Here, we use the solubility of PVP in ethanol, acetone, and isopropyl alcohol (IPA) to treat the ZnO nanofibers and remove the residual PVP in the nanofibers. The J-V characteristics for the hybrid solar cells with surface-treated ZnO nanofibers are shown in **Figure 15**. Corresponding performance parameters are presented in **Table 1**. The J_{sc} of the solar cell decreases dramatically after immersing the ZnO nanofibers into an ethanol solution. This effect is aggravated by an ultrasonic cleaning, which results in a smaller fill factor of 0.27 compared to the immersed device. While the treatment with acetone and IPA exhibits positive effect. After ultrasonic cleaning in acetone and IPA subsequently for 1 min, the open voltage of the solar cells increases from 0.54 to 0.61 V. It implies that the ZnO nanofibers have more contact with the active layer. The contact quality between the electron transport layer like ZnO nanofibers and the active layer like the P3HT:PCBM blend layer can be reflected by the values of shunt resistance (R_{sh}) and series resistance (R_s). Researchers found that contact resistance and charge recombination at the interface are major reasons for giving rise to the increase in R_s and the drop in R_{sh}, respectively [39]. A low R_s of 14 Ω cm^2 and a high R_{sh} of 952 Ω cm^2 strongly support the assertion that ultrasonic cleaning in acetone and IPA could effectively remove the residual PVP from the ZnO nanofibers and ensure a good contact between the ZnO nanofibers and the P3HT:PCBM layer.

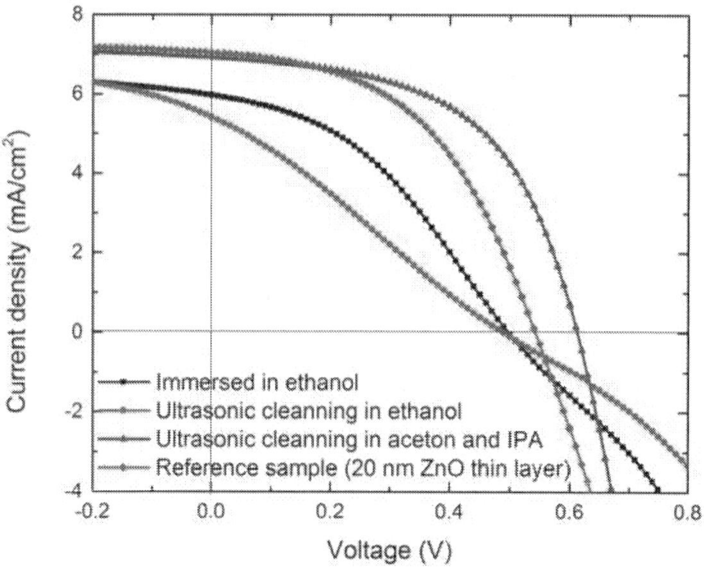

Figure 15. J-V characteristics for hybrid solar cells with ZnO nanofibers after different surface treatment.

Device	Surface treatment method	PCE [%]	V_{oc} [V]	J_{sc} [mA/cm²]	FF	R_{sh} [Ω cm²]	R_s [Ω cm²]
A	Immersed in ethanol	1.18	0.50	5.98	0.40	500	53
B	Ultrasonic cleaning in ethanol	0.72	0.49	5.42	0.27	145	106
C	Ultrasonic cleaning in acetone and IPA	2.31	0.61	6.93	0.54	952	18
D	20 nm ZnO thin layer with no treatment	1.87	0.54	7.04	0.49	869	25

Table 1. Averaged solar cell performance of the hybrid solar cells with surface-treated ZnO nanofibers as electron transport media. PCE: power conversion efficacy; V_{oc}: open-circuit voltage; J_{sc}: short-circuit current; FF: fill factor; R_{sh}: shunt resistance; $_{Rs}$: series resistance.

Although electrospun nanofibers show great potential in hybrid solar cells applications, there are still some problems to be solved. (1) Removing residual polymers. The polymers using for fabricating metal oxide nanofibers should be removed before applied to hybrid solar cells. Only calcination cannot remove completely the residual polymers, which limits the function of metal oxide nanofibers. (2) Reducing the diameter of nanofibers and spacing to around 10 nm. As we discussed in Section 2.2, the diameter of electrospun metal oxide nanofibers is usually in the range of several tens of nanometers (30 nm to several micrometres in case of electrospun ZnO nanofibers), which is still above the diffusion length of exciton. The diffusion length of excitons in electronic devices like solar cells is usually around 10 nm. Hence, electrospun metal oxide nanofibers with finer size are expected to get improved excition dissociation and higher power conversion efficiency. (3) Controlling the orientation of nanofibers. Vertically aligned metal oxides could provide an ideal structure for suppressing of the recombination of dissociated electrons and holes. But most of the orientations of electrospun nanofibers are parallel to the substrate. Vertically aligned electrospun nanofibers have addressed seldom until now. Control of orientation, particularly vertical orientation, is still a challenge for the hybrid solar cell application of electrospun nanofibers. In general, the possibility to control the purity, diameter, spacing, and orientation on metal oxide nanofibers is a critical issue for all researchers.

3.1.2. Dye-sensitized solar cells

Dye-sensitized solar cells usually exploit porous structure of metal oxide nanoparticles to absorb sensitizing dyes, see **Figure 16**. The usage of particles, however, could result in charge-carrier recombination and low power conversion efficiency, because the conduction of electrons in the matrix of nanoparticles could be diminished across the boundaries of the metal oxide nanoparticles.

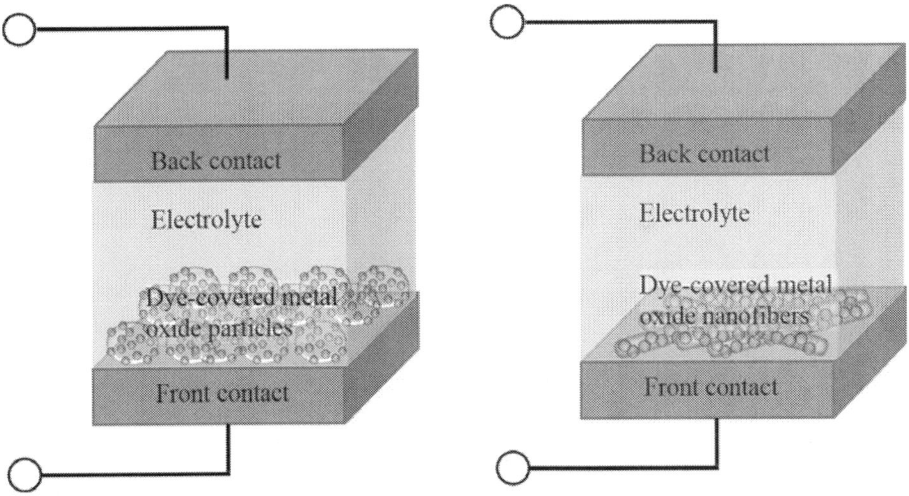

Figure 16. A scheme for dye-sensitized solar cells with dye-covered metal oxide particles (left) and dye-covered metal oxide nanofibers (right).

Comparing with sintered nanoparticles, one-dimensional metal oxide nanofibers can produce high charge conduction because they have smaller particle boundaries and higher specific surface area (see **Figure 16**), which can absorb more sensitizing dyes [40]. The size of the pores in electrospun metal oxide nanofibers is usually larger than that in metal oxide nanoparticles, which facilitates the filling of viscous electrolyte into nanofibers. It has been demonstrated that dye-sensitized solar cells using electrospun TiO_2 nanofibers as electrode could reach power conversion efficiency comparable with that produced by liquid electrolyte system [41]. However, there are problems when applying electrospun mental oxide nanofibers to dye-sensitized solar cells. The porous metal oxide electrodes have poor adhesion to the substrate after removing the polymer component through calcination at high temperature. As shown in **Figure 9**, the metal oxide nanofibers shrink prominently after calcination because of the large stress generated at high temperature. The shrinkage in calcination could affect tremendously not only dye-sensitized solar cell devices but also other energy applications of metal oxide nanofibers. More details about the effect on hybrid solar cell devices are given later. A variety of methods have been reported to solve the adhesion problem for metal oxide nanofibers. Some researchers use pretreatment or solvent vapour to diminish the stress effect during calcination and improve the adhesion between metal oxide nanofibers and the substrate [42].

3.2. Lithium-ion battery applications

Lithium-ion batteries have attracted considerable attention in both research and industrial development due to their advantages, such as high energy density, long cycle life, low gravimetric density, and flexible design. Rechargeable solid lithium-ion batteries have been widely applied to portable electronic devices for more than 25 years. The solid lithium-ion batteries produce electrical energy by converting chemical energy via redox reactions at the electrode. Similarly to working mechanism of fuel cells, lithium-ion batteries comprise an electrolyte sandwiched by an anode and a cathode. The electrolyte is an insulator for electrons but a conductor for Li^+. Carbon/graphite is commonly used anode material for conventional batteries. A lithium-metal oxide (e.g., $LiCoO_2$, $LiMn_2O_4$, and $LiNiO_2$) is usually selected as the cathode materials. And the organic electrolyte could be lithium hexafluorophosphate ($LiPF_6$) salt.

As the demand for energy increases, lithium-ion batteries are expected to be applied to more applications requiring fast charging and discharging at high power rates like electric vehicles. So, many improvements, for instance, cycling life, charge/discharge rate, power, energy density, are needed to meet the requirements. Researchers have developed various novel materials and nanostructures for both electrodes and electrolytes of lithium-ion batteries. Electrospun nanofibers is one of the most promising structures for high-performance lithium-ion batteries due to their one-dimensional structure and low manufacturing cost. $LiCoO_2$ nanofibers have been used as the cathode of lithium-ion battery to get a fast solid state diffusion [43]. While the electrospun $LiCoO_2$ nanofibers have a large loss in capacity during the charge-discharge process, resulting from the dissolution of cobalt and lithium cations forming Li_2CO_3 and CoF_2 impurities. To increase stability and cycling performance of lithium-ion batteries, coatings on the electrospun nanofibers have also been investigated. Besides cathode materials, electrospun nanofibers have been employed for use as anode and electrolyte.

3.3. Other applications

Electrospun metal oxide nanofibers can also be applied to other devices, such as fuel cells, supercapacitors, and so on. For example, many electrospun metal oxide (e.g., TiO_x, MnO_x, WO_x, and SnO_2) nanofibers can be used as catalyst support in fuel cells.

4. Conclusions

The diameter of electrospun metal oxide nanofibers can be controlled by adjusting the properties of the solution. Ultrathin nanofibers formed when using very dilute solution. But fabrication of metal oxide nanofibers at the scale of 10 nm or smaller is still a challenge. Calcination process after electrospinning determined the chemical composition and crystallization of resultant nanofibers. It could help crystallize metal oxide nanofibers, but somewhat limit the practical application of electrospun nanofibers due to the high temperature. Residual polymers in the calcined metal oxide nanofibers could be removed by chemical surface treatments. After surface treatment, the nanofibers have more contact with the active layer, which contributes high performance for solar cells.

Acknowledgements

The authors gratefully acknowledge the support by the JSPS KAKENHI project (No. 26420267). The authors also thank Mr. Suzuki in Saitama University for his help in several experiments.

Author details

Yingjie Liao[1*], Takeshi Fukuda[2] and Suxiao Wang[3]

*Address all correspondence to: ajiewy@gmail.com

1 Centre de Recherches sur les Macromolécules Végétales (CERMAV, UPR-CNRS 5301) affiliated with Grenoble Alpes University, Grenoble, France

2 Department of Functional Materials Science, Saitama University, Saitama, Japan

3 Shanghai Aircraft Design and Research Institute, Shanghai, China

References

[1] Serp P, Corrias M, Kalck P. Carbon nanotubes and nanofibers in catalysis. Appl Catal A-Gen. 2003;253(2):337–58.

[2] Lu XF, Zhang WJ, Wang C, Wen TC, Wei Y. One-dimensional conducting polymer nanocomposites: synthesis, properties and applications. Prog Polym Sci. 2011;36(5): 671–712.

[3] De Jong KP, Geus JW. Carbon nanofibers: catalytic synthesis and applications. Catal Rev-Sci Eng. 2000;42(4):481–510.

[4] Yoshimoto H, Shin YM, Terai H, Vacanti JP. A biodegradable nanofiber scaffold by electrospinning and its potential for bone tissue engineering. Biomaterials. 2003;24(12): 2077–82.

[5] Yang F, Murugan R, Wang S, Ramakrishna S. Electrospinning of nano/micro scale poly(L-lactic acid) aligned fibers and their potential in neural tissue engineering. Biomaterials. 2005;26(15):2603–10.

[6] Sill TJ, von Recum HA. Electro spinning: applications in drug delivery and tissue engineering. Biomaterials. 2008;29(13):1989–2006.

[7] Pham QP, Sharma U, Mikos AG. Electrospinning of polymeric nanofibers for tissue engineering applications: a review. Tissue Engineering. 2006;12(5):1197–211.

[8] Gopal R, Kaur S, Ma ZW, Chan C, Ramakrishna S, Matsuura T. Electrospun nanofibrous filtration membrane. J Membr Sci. 2006;281(1–2):581–6.

[9] Gibson P, Schreuder-Gibson H, Rivin D. Transport properties of porous membranes based on electrospun nanofibers. Colloids Surf A: Physicochem Eng Aspects 2001;187:469–81.

[10] Barhate RS, Ramakrishna S. Nanofibrous filtering media: filtration problems and solutions from tiny materials. J Membr Sci. 2007;296(1–2):1–8.

[11] Liang D, Hsiao BS, Chu B. Functional electrospun nanofibrous scaffolds for biomedical applications. Adv Drug Delivery Rev. 2007;59(14):1392–412.

[12] Burger C, Hsiao BS, Chu B. Nanofibrous materials and their applications. Annu Rev Mater Res. 2006; 36:333–68.

[13] Zong XH, Kim K, Fang DF, Ran SF, Hsiao BS, Chu B. Structure and process relationship of electrospun bioabsorbable nanofiber membranes. Polymer. 2002;43(16):4403–12.

[14] Virji S, Huang JX, Kaner RB, Weiller BH. Polyaniline nanofiber gas sensors: examination of response mechanisms. Nano Lett. 2004;4(3):491–6.

[15] Liu HQ, Kameoka J, Czaplewski DA, Craighead HG. Polymeric nanowire chemical sensor. Nano Lett. 2004;4(4):671–5.

[16] Huang JX, Virji S, Weiller BH, Kaner RB. Polyaniline nanofibers: facile synthesis and chemical sensors. J Am Chem Soc. 2003;125(2):314–5.

[17] Wu Q, Xu YX, Yao ZY, Liu AR, Shi GQ. Supercapacitors based on flexible graphene/polyaniline nanofiber composite films. ACS Nano. 2010;4(4):1963–70.

[18] Ramakrishna S, Fujihara K, Teo WE, Yong T, Ma ZW, Ramaseshan R. Electrospun nanofibers: solving global issues. Mater Today. 2006;9(3):40–50.

[19] Antolini E. Carbon supports for low-temperature fuel cell catalysts. Appl Catal B: Environ. 2009;88(1–2):1–24.

[20] Wu H, Zhang R, Liu XX, Lin DD, Pan W. Electrospinning of Fe, Co, and Ni nanofibers: synthesis, assembly, and magnetic properties. Chem Mater. 2007;19(14):3506–11.

[21] Li D, Herricks T, Xia YN. Magnetic nanofibers of nickel ferrite prepared by electrospinning. Appl Phys Lett. 2003;83(22):4586–8.

[22] Jose R, Thavasi V, Ramakrishna S. Metal oxides for dye-sensitized solar cells. J Am Ceram Soc. 2009;92(2):289–301.

[23] Antolini E, Gonzalez ER. Ceramic materials as supports for low-temperature fuel cell catalysts. Solid State Ionics. 2009;180(9–10):746–63.

[24] Agarwal S, Greiner A, Wendorff JH. Functional materials by electrospinning of polymers. Prog Polym Sci. 2013;38(6):963–91.

[25] Lakshmi BB, Patrissi CJ, Martin CR. Sol–gel template synthesis of semiconductor oxide micro- and nanostructures. Chem Mater. 1997;9(11):2544–50.

[26] Hartgerink JD, Beniash E, Stupp SI. Self-assembly and mineralization of peptide-amphiphile nanofibers. Science. 2001;294(5547):1684–8.

[27] Li D, Xia YN. Electrospinning of nanofibers: reinventing the wheel? Adv Mater (Weinheim, Germany). 2004;16(14):1151–70.

[28] Lee MR, Eckert RD, Forberich K, Dennler G, Brabec CJ, Gaudiana RA. Solar power wires based on organic photovoltaic materials. Science. 2009;324(5924):232–5.

[29] Huynh WU, Peng XG, Alivisatos AP. CdSe nanocrystal rods/poly(3-hexylthiophene) composite photovoltaic devices. Adv Mater (Weinheim, Germany). 1999;11(11): 923–927.

[30] Fan X, Chu ZZ, Wang FZ, Zhang C, Chen L, Tang YW, et al. Wire-shaped flexible dye-sensitized solar cells. Adv Mater (Weinheim, Germany). 2008;20(3):592–595.

[31] Bessel CA, Laubernds K, Rodriguez NM, Baker RTK. Graphite nanofibers as an electrode for fuel cell applications. J Phys Chem B. 2001;105(6):1115–8.

[32] Thavasi V, Singh G, Ramakrishna S. Electrospun nanofibers in energy and environmental applications. Energy Environ Sci. 2008;1(2):205–21.

[33] Liao Y, Fukuda T, Kamata N, Tokunaga M. Diameter control of ultrathin zinc oxide nanofibers synthesized by electrospinning. Nano Res Lett. 2014;9:267.

[34] Zhang X, Thavasi V, Mhaisalkar SG, Ramakrishna S. Novel hollow mesoporous 1D TiO2 nanofibers as photovoltaic and photocatalytic materials. Nanoscale. 2012;4(5): 1707–16.

[35] Gonzalez-Valls I, Lira-Cantu M. Vertically-aligned nanostructures of ZnO for excitonic solar cells: a review. Energy Environ Sci. 2009;2(1):19–34.

[36] Olson DC, Piris J, Collins RT, Shaheen SE, Ginley DS. Hybrid photovoltaic devices of polymer and ZnO nanofiber composites. Thin Solid Films. 2006;496(1):26–9.

[37] Aliaga C, Park JY, Yamada Y, Lee HS, Tsung C-K, Yang P, et al. Sum frequency generation and catalytic reaction studies of the removal of organic capping agents from Pt nanoparticles by UV–ozone treatment. J Phys Chem C. 2009;113(15):6150–5.

[38] Small CE, Chen S, Subbiah J, Amb CM, Tsang SW, Lai TH, et al. High-efficiency inverted dithienogermole-thienopyrrolodione-based polymer solar cells. Nat Photonics. 2012;6(2):115–20.

[39] Liang ZQ, Zhang QF, Wiranwetchayan O, Xi JT, Yang Z, Park K, et al. Effects of the morphology of a ZnO buffer layer on the photovoltaic performance of inverted polymer solar cells. Adv Funct Mater. 2012;22(10):2194–201.

[40] Onozuka K, Ding B, Tsuge Y, Naka T, Yamazaki M, Sugi S, et al. Electrospinning processed nanofibrous TiO2 membranes for photovoltaic applications. Nanotechnology. 2006;17(4):1026–31.

[41] Song MY, Kim DK, Jo SM, Kim DY. Enhancement of the photocurrent generation in dye-sensitized solar cell based on electrospun TiO2 electrode by surface treatment. Synth Met. 2005;155(3):635–8.

[42] Fujihara K, Kumar A, Jose R, Ramakrishna S, Uchida S. Spray deposition of electrospun TiO 2 nanorods for dye-sensitized solar cell. Nanotechnology. 2007;18(36):365709.

[43] Gu Y, Chen D, Jiao X. Synthesis and electrochemical properties of nanostructured LiCoO2 fibers as cathode materials for lithium-ion batteries. J Phys Chem B. 2005;109(38):17901–6.

Chapter 10

Syntheses and Applications of Titanium Compound Nanofiber Thin Films

Mitsunori Yada

Additional information is available at the end of the chapter

http://dx.doi.org/10.5772/64072

Abstract

Syntheses, characteristics, wettability, and antibacterial activity of a titanium phosphate nanobelt thin film and titanate nanofiber thin films formed on titanium plates are presented in this chapter.

Keywords: titanium phosphate, sodium titanate, wettability, antibacterial activity, thin film

1. Introduction

Vigorous efforts have been made in recent years to control the structure and morphology of ceramic nanoparticles. The formation of peculiar nanostructures and morphologies can expose a specific crystal surface or cause a high specific surface area or characteristic pore diameter distribution, which can result in the manifestation of various peculiar physical and chemical properties. Vigorous efforts have been made to form the thin film comprised of nanoparticles, which should help peculiar physical and chemical properties of nanoparticles manifest more efficiently. The dip-coating, spin-coating, and electrophoretic deposition methods are commonly used to form the thin film, but they tend to cause nanoparticles to be randomly deposited on substrates, and the adhesion between the particles and substrate is often not good. An alternative method to form the thin film involves forming a nanostructured metal compound thin film on a metal surface where the metal itself is used as the raw material. Because the metal compound grows crystals directly on the metal surface in the nanostructured metal compound thin film/metal composites, superior adhesion between the metal and the thin film can be expected. In addition, the particle arrangement can be controlled by aligning the crystal orientations of the particles to some degree. The superior electrical conductivity and mechanical properties of the metal and the peculiar physical and chemical properties of the nanostructured metal compound thin film can potentially provide a superior functional material that combines all such properties. Vigorous research efforts have been made on this aspect, particularly with regard to the synthesis of a nanostructured titanium compound thin film/titanium composite. This is because titanium compounds such as titanium dioxide and titanate feature superior characteristics for a broad range of applications, e.g., photocatalysts [1], dye-sensitized solar cells [2], lithium-ion batteries [3, 4], and biomaterials [5, 6]. Titanium dioxide and titanate are also extremely attractive because they are available at relatively low prices and are safe materials with no toxicity. Titanium dioxide and titanate have reportedly been used in such syntheses as the formation of a titanium dioxide nanotube thin film on a titanium metal surface with an electrochemical method (anodic oxidation of titanium metal) [6, 7], the formation of nanostructured titanate (nanotube, nanofiber, nanosheet, etc.) thin films on titanium metal surfaces by the reaction of titanium metal in an alkaline aqueous solution, and their conversions into nanostructured titanium dioxide thin film [5, 8–11]. For example, known applications of composites that utilize the superior electrical conductivity of titanium include electrode materials of dye-sensitized solar cells [12], lithium-ion batteries [13], and sodium-ion batteries [14]. Superior electrical conductivity can be secured by using titanium dioxide or titanate with one-dimensional morphologies, such as nanorods, nanotubes, or nanofibers, while electrodes can be prepared without using binders or conducting agents. Applied research on field emission materials [15] and electrochromic materials [16] of composites is also underway. On the other hand, biomaterials are a known application that utilizes the superior mechanical properties and safety of titanium metal. Titanium and titanium alloys are used as artificial joints based on their mechanical properties and safety inside the living body. Using composites adds osteoconductivity derived from titanium dioxide or titanate and antibacterial properties to titanium and titanium alloys, which facilitates the safe adhesion of bones and artificial joints in a short period of time [5, 6]. Antibacterial biomaterials are also desirable for medical sites; infections occur in 1–2% of artificial joint replacement surgeries and are a major problem [17, 18]. The present chapter evaluated the synthesis and antibacterial properties of titanate nanofiber thin film/titanium metal composites as an example application for biomaterials [10]. Various crystal structures such as $Ti(HPO_4)_2{\cdot}H_2O$ and $Ti(H_2PO_4)PO_4{\cdot}2H_2O$ are known to exist with titanium phosphate. Morphology controls to shape nanotubes [19], nanofibers [20], elongated hexagonal nanoplates [21], and so on are also being developed. Applications of titanium phosphate include not just fundamental research on topics such as ion exchange and intercalation reactions but also catalysts [22], selective absorbents for toxic substances (e.g., Pb^{2+}, Cd^{2+}) [23, 24], the support and release of drugs [25], sodium batteries [26], and solar cells [27]. However, examples of reports on such efforts are primarily about particles, and there have not been many examples of studies on thin films. In particular, there are only three reports on the synthesis of titanium phosphate thin film on the surface of titanium, including the research of the present author. Lu et al. [28] immersed titanium in a phosphoric acid aqueous solution and induced a hydrothermal reaction at 250°C, while Park et al. [29] also immersed titanium in a phosphoric acid aqueous solution and induced a hydrothermal reaction at 180°C to form $Ti_2O(PO_4)_2(H_2O)_2$ nanorod thin films on the titanium surface. Park et al. [29] revealed that such a thin film has

http://dx.doi.org/10.5772/64072

superior osteoconductivity. The present author added to these prior studies by discovering that titanium phosphate featuring a variety of crystal structures and morphologies is generated on the surface of a titanium plate immersed in an aqueous solution of hydrogen peroxide and phosphoric acid to produce a hydrothermal reaction [30]. This book is focused on nanofibers; this chapter narrows the focus to the titanium phosphate nanobelt thin film/titanium metal composite by introducing the formation mechanism and wettability.

2. Synthesis and wettability of a titanium phosphate nanobelt thin film

A titanium plate (20 mm × 20 mm × 0.5 mm) was scoured by using an abrasive paper and then immersed in 10 mL of an aqueous solution of 2 mol/L hydrogen fluoride for 10 min at room temperature to remove the surface-oxidized phase. It was then repeatedly washed with distilled water. A synthesis experiment of titanium phosphate thin film was performed with the derived plate. This experimental manipulation was performed in order to obtain a homogeneous titanium surface. This manipulation changed the morphology of the surface from that shown in **Figure 1a** to that shown in **Figure 1b** with a homogeneous surface morphology derived for each experiment. This plate was immersed in a mixture comprising 9.2 mL of 30 wt% H_2O_2 solution and 0.5 g of 85 wt% H_3PO_4 solution and left to react for 24 h at 120°C. Depending on the synthesis conditions (e.g., the raw material ratio or reaction temperature), this reaction generates high pressure inside the reaction container due to the generated gas, particularly when the hydrothermal reaction takes place at 100°C or higher. A reactor vessel that can measure the internal gas pressure while a reaction occurs inside and that is equipped with a safety valve to release the gas when required is desired. The obtained product was repeatedly washed with distilled water after the reaction. Because titanium phosphate particles that were generated in the solution during reaction deposited on the upper surface (front face), it was not possible to accurately evaluate the morphology of the titanium phosphate thin film growing from the titanium plate. Thus, a detailed evaluation was performed on the lower surface (back face) on which particles did not deposit. A thin film formed evenly on the surface of the titanium plate, as shown in **Figure 2a**. The thin film is composed of nanobelts with approximate widths of 30–270 nm and lengths of a few microns or more formed on the surface of the titanium plate, as shown in **Figure 2b** and **c**. The TEM image (**Figure 2c**) confirmed that the particles constituting the thin film were nanobelts because they were thin one-dimensional in morphology. Diffraction peaks that were attributed to titanium metal and $Ti_2O_3\ (H_2PO_4)_2{\cdot}2H_2O$ were observed in the X-ray diffraction measurements (**Figure 3**). They revealed that the nanobelts were made of $Ti_2O_3\ (H_2PO_4)_2{\cdot}2H_2O$. Korösi et al. [25] reported on rectangular $Ti_2O_3(H_2PO_4)_2{\cdot}2H_2O$ nanoparticles with dimensions of 15 ± 3 nm × 20–150 nm. The nanobelts observed during this experiment, however, had nanoparticles with significant growth in the long axis direction. The thin film thickness was about 8 μm at the reaction time of 24 h but this thin film thickness became broader and reached 20 μm when the reaction time was extended to 72 h. SEM image of the thin film cross-section at the reaction time of 72 h are shown in **Figure 2d**. A phase comprising nanobelts was observed from the top surface of the thin film down to a depth of about 18 μm. On the side closer to the titanium plate of the thin film, however, a phase of about 2 μm that was relatively dense was observed.

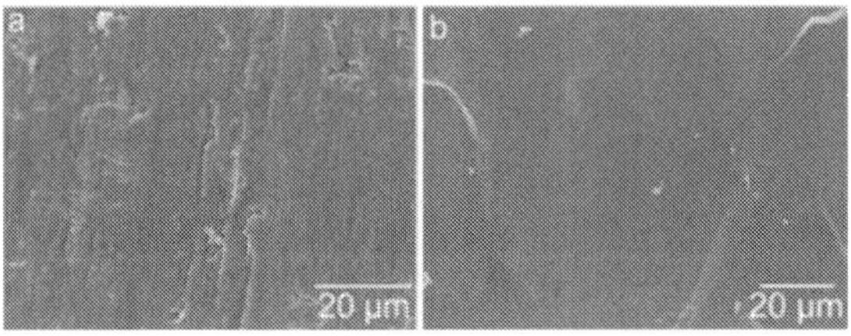

Figure 1. SEM images of the titanium plates before (a) and after (b) the HF treatment.

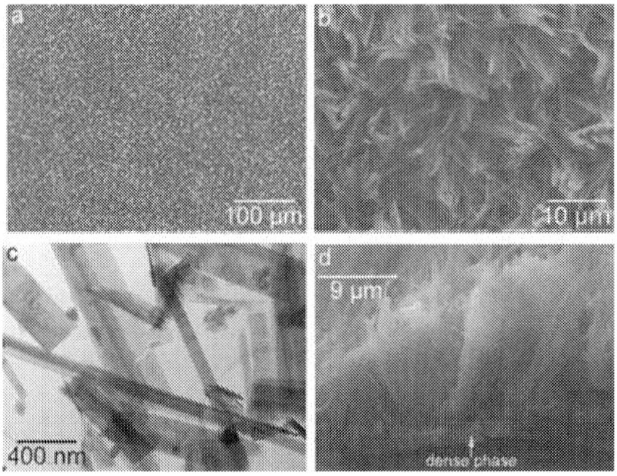

Figure 2. SEM images (a, b, and d) and a TEM image (c) of the titanium phosphate nanobelt thin films formed on titanium plates.

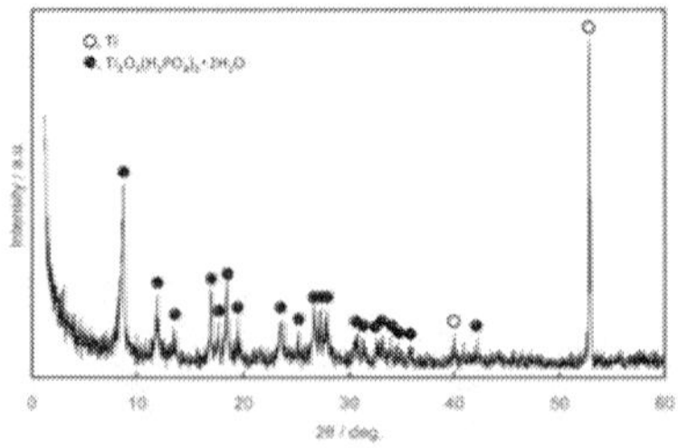

Figure 3. XRD pattern of the titanium phosphate nanobelt thin film.

The role of hydrogen peroxide was examined in order to consider the formation mechanism of nanobelts. This is because past studies that did not involve the use of hydrogen peroxide did not generate Ti_2O_3 $(H_2PO_4)_2{\cdot}2H_2O$, as described earlier [28, 29]. In order to investigate the impact of hydrogen peroxide on the formation of the thin film, synthesis was performed with a fixed reaction temperature, reaction time, and phosphoric acid quantity of 120°C, 24 h, and 0.5 g, respectively, with varying quantities of hydrogen peroxide and water. When 9.2 mL of water was used for synthesis instead of hydrogen peroxide, the thin film did not form, and the formation of TiH_2 was observed in the XRD measurement. This confirmed that hydrogen peroxide plays an important role in the formation of the $Ti_2O_3(H_2PO_4)_2{\cdot}2H_2O$ nanobelt thin film. **Figure 4** shows a schematic representation of the formation mechanism for nanostructured titanium phosphate thin films based on the above results. Hydrogen peroxide has two primary roles. First, it acts as an oxidizer. Hydrogen peroxide acts as an oxidizer in acid solutions according to the following reaction formula:

$$H_2O_2 + 2H^+ + 2e^- \rightarrow 2H_2O$$

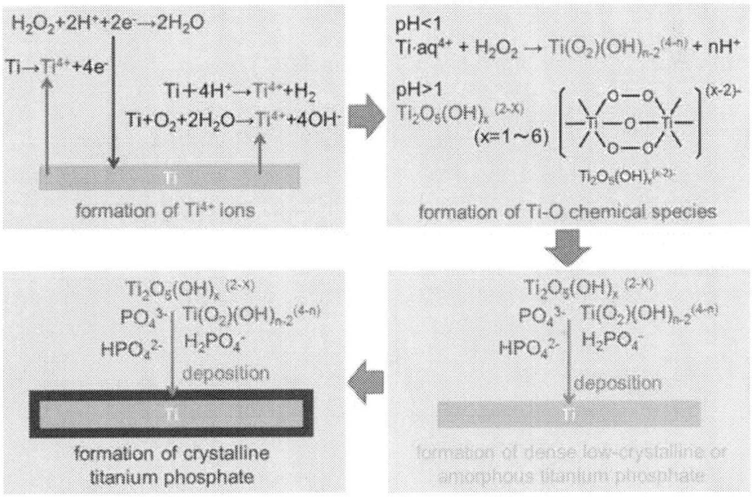

Figure 4. Schematic representation of the formation mechanism of the nanostructured titanium phosphate thin film formed on a titanium plate.

Therefore, hydrogen peroxide promotes the generation of Ti^{4+} from the titanium plate, as described by the following equation:

$$Ti + 2H_2O_2 + 4H^+ \rightarrow Ti^{4+} + 4H_2O \ \ (H_2O_2 + 2H^+ + 2e^- \rightarrow 2H_2O, \ Ti \rightarrow Ti^{4+} + 4e^-)$$

Although Ti^{4+} can be generated according to the following reaction formula by H^+ or dissolved oxygen in the aqueous solution, hydrogen peroxide has the greatest effect on promoting Ti^{4+} generation:

$$Ti + 4H^+ \rightarrow Ti^{4+} + H_2$$

$$Ti + O_2 + 2H_2O \rightarrow Ti^{4+} + 4OH^-$$

Ti^{4+} eluted from Ti into the reaction solution immediately starts to react with H_2O_2. Ti^{4+} is known to react with hydrogen peroxide when pH <1 [31].

$$Ti{\cdot}aq^{4+} + H_2O_2 \rightarrow Ti(O_2)(OH)_{n-2}{}^{(4-n)+} + nH^+$$

Furthermore, binuclear complexes such as $Ti_2O_5\,(OH)_x{}^{(2-X)}$ (x = 1–6) are known to be generated when pH >1 [31]. These are referred to as peroxotitanic acid complexes and are indicated by a yellow color in the aqueous solution. The color of the solution after synthesis of the titanium phosphate nanobelt thin film was actually yellow, as shown in **Figure 5**, which indicates that a peroxotitanic acid complex was formed. Therefore, the second role of hydrogen peroxide is the formation of chemical species that contain Ti–O bonds. When such chemical species react with $H_2PO_4^-$ ions, a relatively dense phase is formed first as an amorphous or low-crystalline titanium phosphate phase precipitates on the titanium plate. The nanobelt thin film forms next as high-crystalline $Ti_2O_3(H_2PO_4)_2{\cdot}2H_2O$, which is the most stable phase, grows on the surface of the dense phase. The formation of such a dense amorphous phase is important to link the titanium metal with $Ti_2O_3(H_2PO_4)_2{\cdot}2H_2O$, which has a completely different crystal structure. The phenomenon of a peculiar nanostructured thin film forming via the formation of a relatively dense phase on a metal surface has also been observed for the formation of sodium titanate thin films on titanium plate surfaces [9], which is described later. Synthesis using a mixture comprising 2.3 mL of hydrogen peroxide and 6.9 mL of water instead of 9.2 mL of hydrogen peroxide led to a flower-like aggregate of semispherical morphologies having a diameter of about a few tens of microns while the composition remained the same as $Ti_2O_3(H_2PO_4)_2{\cdot}2H_2O$. This means that a thin film comprising microflowers was generated (**Figure 6a**). Observation of the cross-section revealed that the thickness of the thin film was about 30 μm, and the microflower comprised particles with a nanoplate morphology about 400–700 nm wide and about 30–40 μm long growing in a radial pattern. A thin film of $Ti_2O_3(H_2PO_4)_2{\cdot}2H_2O$ having an intermediate morphology between the nanobelt thin film (**Figure 2b**) and microflower thin film (**Figure 6a**) was formed by using a mixture comprising 4.6 mL of hydrogen peroxide and 4.6 mL of water instead of 9.2 mL of hydrogen peroxide, as shown in **Figure 6b**. These results indicate the following formation process. The number of nuclei of $Ti_2O_3(H_2PO_4)_2{\cdot}2H_2O$ formed on the titanium plate decreased significantly once the amount of hydrogen peroxide in the reaction solution decreased. The microflower thin film was believed to have been formed as a large number of $Ti_2O_3(H_2PO_4)_2{\cdot}2H_2O$ crystals slowly grew radially from these nuclei, which were just a few in number. Therefore, the morphology of the thin film can clearly be controlled in a variety of ways while the composition of $Ti_2O_3(H_2PO_4)_2{\cdot}2H_2O$ is maintained by controlling the synthesis conditions as described above.

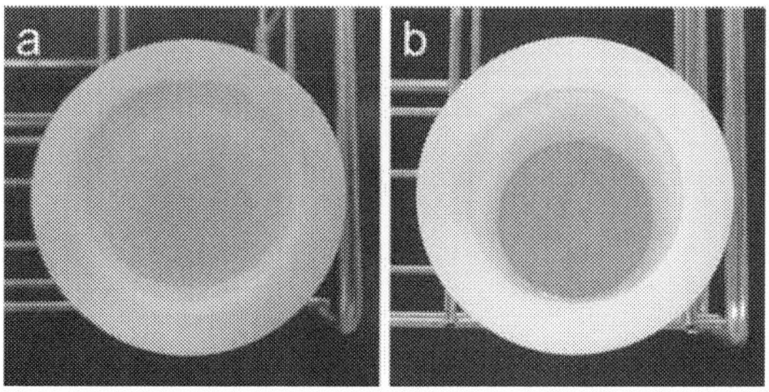

Figure 5. Digital camera images of the solutions before (a) and after (b) the reaction.

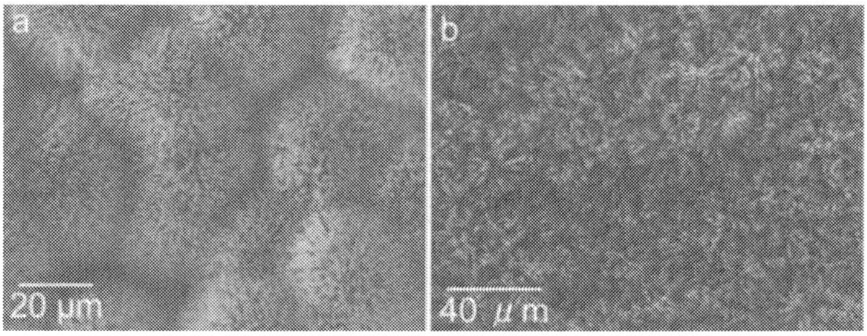

Figure 6. SEM images of $Ti_2O_3(H_2PO_4)_2{\cdot}2H_2O$ thin films formed on titanium plates.

Thin films with compositions other than $Ti_2O_3(H_2PO_4)_22H_2O$ or of various morphologies other than nanobelts can be generated by controlling the raw material ratios and reaction temperatures, although no detailed descriptions are provided here because this chapter is focused on nanobelts. For instance, a thin film composed of anatase-type titanium dioxide and about 2 μm thick, a few tens of nanometers wide, and about 2 μm long—similar to that reported by Wu et al. [32]—was generated when synthesis was performed with a reaction temperature of 120°C, 9.2 mL of 30 wt% H_2O_2 solution, 0 g of 85 wt% H_3PO_4 solution, and reaction time of 24 h. A thin film about 8 μm thick, which appears to have been composed of hexagonal $Ti(HPO_4)_2{\cdot}H_2O$ nanoplates a few microns in size, was derived when synthesis was performed with a reaction temperature of 120°C, 9.2 mL of 30 wt% H_2O_2 solution, 5 g of 85 wt% H_3PO_4 solution, and reaction time of 24 h (**Figure 7a**). A 2-μm-thick thin film composed of low-crystalline $Ti_2O_3(H_2PO_4)_22H_2O$ nanosheets about a few hundred nanometers to 2 μm in size was formed when synthesis was performed with a reaction temperature of 80°C, 9.2 mL of 30 wt% H_2O_2 solution, 0.5 g of 85 wt% H_3PO_4 solution, and reaction time of 24 h (**Figure 7b**). A thin film about 20 μm thick composed of $Ti_2O(PO_4)_2H_2O$ nanorods about 80–200 nm in diameter and a dozen or so microns long was formed when synthesis was performed with a reaction temperature of 160°C, 9.2 mL of 30 wt% H_2O_2 solution, 0.5 g of 85 wt% H_3PO_4 solution, and reaction time of 24 h (**Figure 7c**).

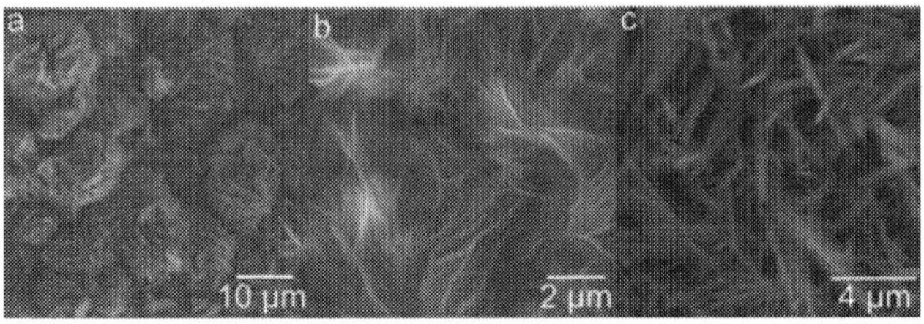

Figure 7. SEM images of the hexagonal $Ti(HPO_4)_2{\cdot}H_2O$ nanoplate thin film (a), the $Ti_2O_3(H_2PO_4)_22H_2O$ nanosheet thin film (b), and the $Ti_2O(PO_4)_2H_2O$ nanorod thin film (c).

The wettability of these thin films was investigated since titanium phosphate thin films having a diverse range of surface morphologies were derived, as described above. The wettability of a surface is generally determined by the surface energy and uneven structure of the surface [33]. In cases where the surface energy is high, the thin film tends to be hydrophilic; when the surface energy is low, it tends to be hydrophobic. Furthermore, the wettability is enhanced when an uneven structure is formed on a surface. When a water droplet was dropped on a synthesized titanium phosphate thin film, the droplet spread instantaneously, which indicates super-hydrophilicity. When the thin film was immersed in a 0.1 mol/L n-dodecyl decylamine hydrochloride solution of 2-propanol, on the other hand, the hydrophilicity decreased because the surface energy decreases when dodecylamine is adsorbed on the surface of titanium phosphate. The contact angle was 25° with the nanobelt thin film, 98° with the microflower thin film, 151° with the nanorod thin film, 151° with the nanoplate thin film, and 128° with the nanosheet thin film. Thus, the contact angle varies significantly depending on the composition and morphology of the thin film. Common features of test specimens with larger contact angles (nanorod thin film, nanoplate thin film and nanosheet thin film) were that the gap between protruding parts was narrow, and the spacing was uniform. With the nanobelt thin film, on the other hand, there were portions where the spacing between protruding parts was narrow and others where they were wide since the nanobelts were bent. The wider sections in such instances decreased the contact angle. Furthermore, the contact angle may have been reduced because there were portions on the microflower thin film where titanium phosphate micro-flowers were not generated and water was absorbed.

3. Synthesis and antibacterial activity of titanate nanofiber thin films

A titanium plate (20 mm × 20 mm × 2 mm) was immersed in 20 mL of a NaOH aqueous solution in a Teflon container; a hydrothermal reaction occurred at 160°C for 20 h. The derived product was then washed repeatedly with water to produce a titanium plate covered by a sodium titanate nanofiber thin film on the surface, i.e., a sodium titanate nanofiber thin film/titanium metal composite. A digital camera photograph and a SEM image (**Figure 8a**) confirmed that nanofibers were formed uniformly on the surface of the titanium metal. SEM observation of the cross-section revealed that the thin film thickness was about 5 μm. Furthermore, the TEM images revealed that the nanofibers were a few tens to a hundred nanometers in width and a few microns in length (**Figure 8b**). The diffraction peaks attributed to titanate were observed in the X-ray diffraction pattern of the thin film. A Na:Ti:O molar ratio of 1:3.04:6.08 was determined from the EDS analysis of the thin film. This approximates the molar ratio for $Na_2Ti_6O_{13}$. Thus, the thin film was considered to be a $Na_2Ti_6O_{13}$ nanofiber thin film. The morphology of the sodium titanate thin film can be controlled according to the concentration and reaction temperature of the aqueous solution of NaOH. Using 1 mol/L NaOH aqueous solution instead of 4 mol/L NaOH aqueous solution resulted in the generation of a nanosheet thin film (**Figure 9a**) [10]. Using 10 mol/L NaOH aqueous solution resulted in the generation of a nanotube thin film (**Figure 9b**) [9]. Using 5 mol/L NaOH aqueous solution for a reaction at 60°C for 24 h resulted in the generation of a thin film with a porous network structure [5, 34].

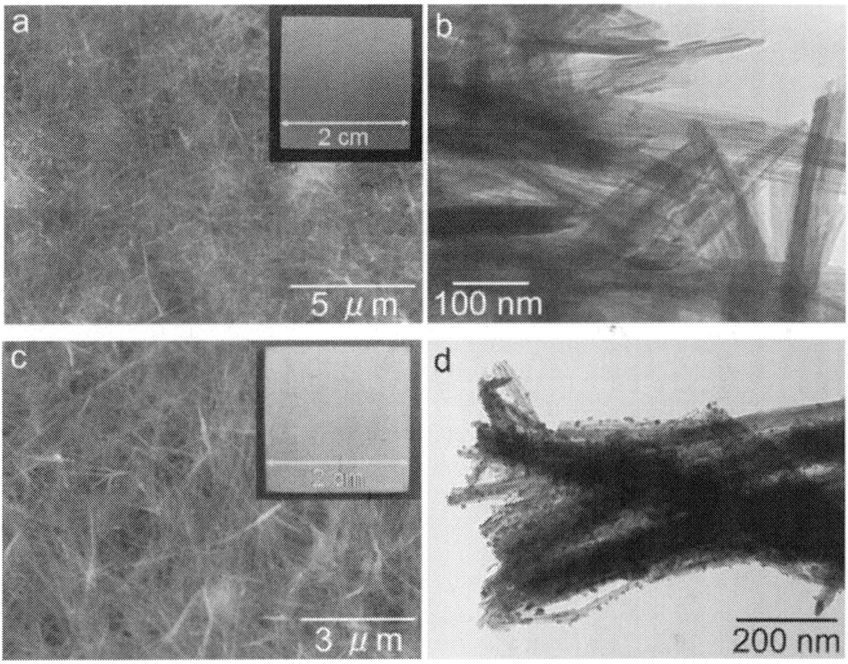

Figure 8. SEM images (a and c) and TEM images (b and d) of the sodium titanate nanofiber thin film (a and b) and the silver nanoparticles/silver titanate nanofiber thin film (c and d). Insets are photographs taken with a digital camera [10].

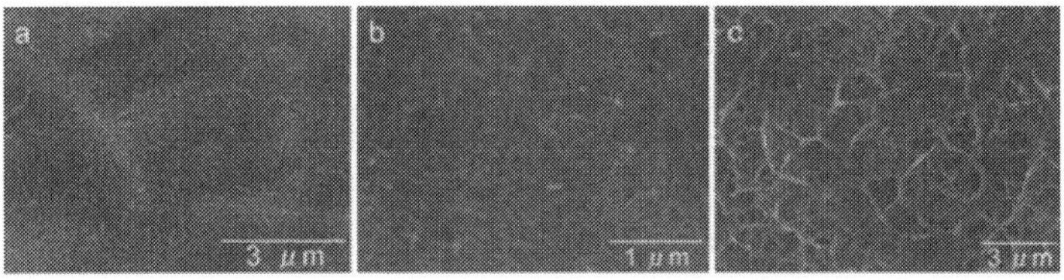

Figure 9. SEM images of the sodium titanate nanosheet thin film (a), the sodium titanate nanotube thin film (b) [34], and the sodium titanate thin film with a porous network structure (c) [34] formed on titanium plates.

The antibacterial activity of sodium titanate nanofiber thin film against methicillin-resistant staphylococcus aureus (MRSA) was investigated. MRSA is one of principal causative agents for nosocomial infection. Two types of tests were performed for the antibacterial test: the test specimen was irradiated and not irradiated with ultraviolet rays with the expectation of photocatalytic activity. Refer to the authors' previous paper for details on the methods used in the antibacterial tests [34]. The antibacterial activity value was used as an index for evaluation. The antibacterial activity value R is defined by the following equation; the antibacterial activity is considered to exist when $R \geq 2$.

$$R = \{\log(B/A) - \log(C/A)\} = \log(B/C)$$

Here, A is the viable bacterial count immediately after inoculation, B is the viable bacterial count of a blank test specimen 24 h later, and C is the viable bacterial count of the nanostructured titanate thin film 24 h later.

A high antibacterial activity value of 5.9 was recorded during the first round of antibacterial tests conducted under exposure to ultraviolet irradiation. Surprisingly, an extremely high antibacterial activity value of 5.8 was recorded even when a similar antibacterial test was conducted with no exposure to ultraviolet rays. This means that a high level of antibacterial activity was recorded for the nanofiber thin film during the first round of antibacterial tests regardless of whether or not the test specimen was exposed to ultraviolet rays. The antibacterial test was repeatedly performed using the same test specimen with and without (i.e., in a dark place) exposure to ultraviolet rays. Extremely high values of 7.7 and 6.6 were recorded for antibacterial activity during the second and third rounds of tests conducted under exposure to ultraviolet rays. Extremely low values of 1.0 and 0.3 were recorded for antibacterial activity during the second and third rounds of antibacterial tests conducted without exposure to ultraviolet rays. The above results revealed that (1) a high level of antibacterial activity manifested with the sodium titanate nanofiber thin film due to photocatalysis effects with exposure to ultraviolet rays and that (2) the sodium titanate nanofiber thin film has a high level of antibacterial activity immediately after synthesis, even when they are not exposed to any ultraviolet rays. Only the sodium titanate nanofiber thin film indicating a high level of antibacterial activity is extremely peculiar because the antibacterial activity of the sodium titanate nanosheet thin film and nanotube thin film described above were 2.1 [10] and 1.7 [34], respectively. Treating such sodium titanate nanofiber thin film with a hydrochloric acid aqueous solution or heat-treating it at 600°C reduced the antibacterial activity value to 0.9 and 1.3, respectively, which eliminated the antibacterial activity. Because bacteria are known to have different properties that classify them as gram-positive or gram-negative depending on their structures, antibacterial tests regarding such individual bacteria types were also conducted. The antibacterial activity value of the sodium titanate nanofiber thin film was 5.5 against *Escherichia coli,* which is a gram-positive bacteria, and 4.7 against *Staphylococcus aureus,* which is a gram-negative bacteria. This reveals that sodium titanate nanofibers have a high level of antibacterial activity against bacteria of both structures. Because the reaction time was extended from 1 to 20 h, the formation of nanofibers was clarified according to the SEM images depicted in **Figure 10**. The intensities of the respective diffraction peaks were greater in the X-ray diffraction pattern, as shown in **Figure 11**. The antibacterial activity values of the thin films generated with reaction times of 1 and 5 h were 1.2 and 0.3, respectively, which indicates no antibacterial activity. The antibacterial activity value for the thin film generated with a reaction time of 12 h was 3.1; thus, the antibacterial activity manifested. The value was 6.5 for the thin film generated with a reaction time of 20 h; thus, a high level of antibacterial activity occurred. The above results indicate that the antibacterial activity increases with the growth of nanofibers and enhanced crystallinity. While it is not possible to offer any clear reason for such high levels of antibacterial activity at present, the following assumptions were made. All nanostructure morphologies (i.e., nanofiber, nanosheet, and nanotube) involve layering and rolling of titanate sheets, and the surface and the edge sections of titanate sheets can be considered separately because of their different chemical properties [35]. Nanofiber structures are characterized by having the highest proportion of the edge section to the overall area compared with other nanostructures. This may mean that the edges of titanate sheets become active sites and affect bacteria in some way, which manifests as a high level of antibacterial activity.

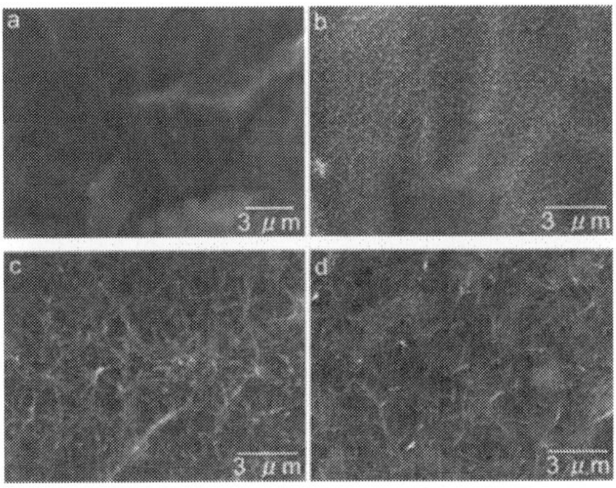

Figure 10. SEM images of sodium titanate nanofiber thin films synthesized following different reaction times; the titanium plate before the synthesis (a), and after 1 h (b), 5 h (c), and 12 h (d) reaction times [10].

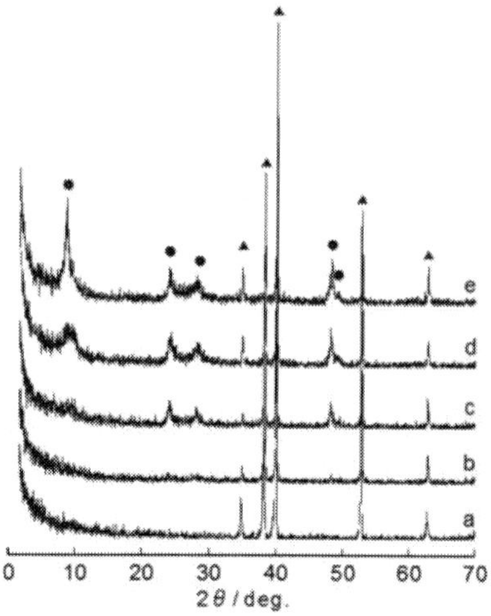

Figure 11. XRD patterns of sodium titanate nanofiber thin films following different reaction times; the titanium plate before the synthesis (a), and after 1 h (b), 5 h (c), 12 h (d), and 20 h (e) reaction times. Peak assignment: ●, sodium titanate; ▲, titanium [10].

Sodium titanate nanofiber thin film is considered to have poor practicality at the current point in time; while superior antibacterial activity was indicated during the first round of tests, the results from the second and subsequent rounds were not good, and there are many unknown aspects about the antibacterial mechanism. Silver, which is already being used in many studies on antibacterial materials, was therefore considered as an antibacterial component based on three considerations: (1) the superior antibacterial activity, (2) some degree of safety for living organisms, and (3) the relative lack of reports regarding the manifestation of drug-resistant bacteria [36]. Sodium titanate is a cation exchanger with a layered structure and can support silver in the titanate by exchanging Na^+ in sodium titanate with Ag^+. A composite treated with silver ion exchange was derived by immersing the sodium titanate nanofiber thin film/titanium metal composite generated with a reaction time of 20 h in 12 mL of a 0.05 mol/L silver acetate aqueous solution at 40°C for 3 h, which was then washed repeatedly with distilled water and dried in a cool and dark location. The thin film derived after the silver ion exchange process remained firmly attached to the titanium plate, as shown in the digital camera photographs and SEM images in **Figure 8c**. There were no changes in the morphology, which had dimensions on the scale of microns. The TEM images, however, revealed precipitation of nanoparticles with dimensions of a few nanometers to a few tens of nanometers on the surface of the nanofiber (**Figure 8d**). These particles were considered to be silver nanoparticles, but no diffraction peaks attributed to silver were observed in the X-ray diffraction pattern (**Figure 12**). This may have been due to the small amount of precipitation that occurred because of the small diameters of the silver nanoparticles. No sodium was detected from the thin film after the silver ion exchange process, but silver was detected when the EDS analysis was conducted. A silver-to-titanium molar ratio of 0.638 was detected; this is greater than the sodium-to-titanium molar ratio of 0.329 before the silver ion exchange process. Such results of the EDS analysis correspond to the precipitation of silver particles on the surface of nanofiber when Na^+ in the sodium titanate nanofiber was exchanged with Ag^+. The X-ray diffraction pattern after the silver ion exchange process indicates the disappearance of the diffraction peak of $2\theta = 8.78°$, which was caused by the layered structure observed with sodium titanate (**Figure 12**). Significant changes in the position and strength of the diffraction peaks at $2\theta \geq 15°$owing to the crystal structure of titanate were observed compared to before the silver ion exchange process (**Figure 12**). Such results indicate that the crystal structure of titanate changed from a two-dimensional layered structure to a three-dimensional crystal structure. The above results reveal that silver nanoparticles and silver titanate nanofiber thin film are generated by the silver ion exchange process of the sodium titanate nanofiber thin film. Such precipitation of silver nanoparticles on the surface of titanate and change in crystal structure caused by the silver ion exchange process have also been observed with the sodium titanate nanotube thin film and the sodium titanate thin film with a porous network structure [34]. This suggests that these are reaction behaviors common to layered titanates.

Figure 12. XRD patterns of the thin film before (a) and after (b) the silver ion exchange treatment. Peak assignment: ●, sodium titanate; ▲, titanium; ○, silver titanate [10].

The antibacterial activity value of the silver nanoparticles/silver titanate nanofiber thin film with respect to MRSA is 7.9, which is extremely high. This value is greater than the antibacterial activity value of 6.3 for a silver nanoparticles/silver titanate nanotube thin film or 7.8 for a silver nanoparticles/silver titanate thin film with a porous network structure [34]. The elution behavior of silver was investigated in fetal bovine serum (FBS) to simulate the environment of a living body because it was presumed to impact the antibacterial activity in a significant manner. Please refer to the authors' previous paper for details on the experimentation methods. Elution of 110 ppm was observed with silver nanoparticles/silver titanate nanofiber thin film, 94 ppm was observed with the silver nanoparticles/silver titanate nanotube thin film, and 68 ppm was observed with the silver particles/silver titanate thin film with a porous network structure on the first day. The greatest amount of silver was clearly eluted by silver nanoparticles/silver titanate nanofiber thin film with the highest level of antibacterial activity. The orders of antibacterial activity and elution quantity for silver did not match completely. Therefore, antibacterial activity cannot be explained solely by the elution amount of silver. Although a clear examination is not possible at the present point in time, the surface morphology of the thin film is considered to be somehow related to the antibacterial activity. Chronologically, the amount of silver elution was 110 ppm on the first day, decreased to 77 ppm by the second day, and gradually decreased with subsequent days to 20 ppm by the tenth day. The elution of silver from silver nanoparticles and from silver titanate nanofiber due to the ion exchange reaction were considered as sources of silver elution from the thin film. FBS is known to contain a large amount of proteins and cations that readily bond with Ag^+, and the ion exchange reaction is known to be extremely rapid. This means that it is extremely likely that the large amount of silver elution on the first day was due to the silver elution from the ion exchange reaction of silver titanate nanofibers. The contribution to silver elution by the silver titanate nanofibers decreased with time; instead, the silver elution from silver nanoparticles increased in their place. The antibacterial activity can be expected to be sustained over a long period since the silver nanoparticles/silver titanate nanofiber thin film can continue to elute silver in this manner for over 10 days.

4. Conclusion

This chapter introduces the generation of titanium compounds with various nanostructures, including titanium phosphate nanobelts and titanate nanofibers on the surface of titanium metal plate, from the treatment of a titanium metal plate in aqueous solutions of various compositions. Because aqueous solutions of various compositions and concentrations and different reaction temperatures and times can be combined for chemical processes involving titanium metal, novel titanium compounds/titanium metal composites with various compositions and nanostructures are expected to be synthesized in the future. It is important to continue with applied research on developing composites that feature the characteristics of both nanostructured titanium compounds and titanium metal, as described in Section "Introduction." In this manner, characteristics that surpass those of the nanostructured titanium compound particles can be derived. Research is expected to continue with applications including biomaterials and electrode materials intended for dye-sensitized solar cells, lithium-ion batteries, and sodium-ion batteries.

Acknowledgements

This research was partially supported by KAKENHI (19750172, 25420732). The author gives special thanks to Dr. Yuko Inoue, Miss Ayako Sakamoto, Miss Misaki Maeda, and Mr. Tomohiro Morita for their supports.

Author details

Mitsunori Yada

Address all correspondence to: yada@cc.saga-u.ac.jp

Department of Chemistry and Applied Chemistry, Faculty of Science and Engineering, Saga University, Saga, Japan

References

[1] Schneider J, Matsuoka M, Takeuchi M, Zhang J, Horiuchi Y, Anpo M, Bahnemann D W. Understanding TiO_2 photocatalysis: mechanisms and materials. Chem Rev. 2014;114:9919–9986.

[2] Roose B, Pathak S, Steiner U. Doping of TiO_2 for sensitized solar cells. Chem Soc Rev. 2015;44:8326–8349.

[3] Sandhya C P, John B, Gouri C. Lithium titanate as anode material for lithium-ion cells: a review. Ionics. 2014;20:601–620.

[4] Song T, Paik U. TiO_2 as an active or supplemental material for lithium batteries. J Mater Chem A. 2016;4:14–31.

[5] Kokubo T, Yamaguchi S. Bioactive Ti metal and its alloys prepared by chemical treatments: state-of-the-art and future trends. Adv Eng Mater. 2010;12:B579–B591.

[6] Kulkarni M, Mazare A, Gongadze E, Perutkova Š, Kralj-Iglič V, Milošev I, Schmuki P, Iglič A, Mozetič M. Titanium nanostructures for biomedical applications. Nanotechnology. 2015;26:062002.

[7] Roy P, Berger S, Schmuki P. TiO_2 nanotubes: synthesis and applications. Angew Chem Int Ed. 2011;50:2904–2939.

[8] Peng X, Chen A. Large-scale synthesis and characterization of TiO_2-based nanostructures on Ti substrates. Adv Funct Mater. 2006;16:1335–1362.

[9] Yada M, Inoue Y, Uota M, Torikai T, Watari T, Noda I, Hotokebuchi T. Plate, wire, mesh, microsphere, and microtube composed of sodium titanate nanotubes on a titanium metal template. Langmuir. 2007;23:2815–2823.

[10] Yada M, Inoue Y, Noda I, Morita T, Torikai T, Watari T, Hotokebuchi T. Antibacterial properties of titanate nanofiber thin films formed on a titanium plate. J Nanomater. 2013:47685.

[11] Inoue Y, Noda I, Torikai T, Watari T, Hotokebuchi T, Yada M. TiO_2 nanotube, nanowire, and rhomboid-shaped particle thin films fixed on a titanium metal plate. J Solid State Chem. 2010;183:57–64.

[12] Wu W-Q, Rao H-S, Xu Y-F, Wang Y-F, Su C-Y, Kuang D-B. Hierarchical oriented anatase TiO_2 nanostructure arrays on flexible substrate for efficient dye-sensitized solar cells. Sci Rep. 2013;3:1892.

[13] Shen L, Uchaker E, Zhang X, Cao G. Hydrogenated $Li_4Ti_5O_{12}$ nanowire arrays for high rate Lithium Ion batteries. Adv Mater. 2012;24:6502–6506.

[14] Li H, Fei H, Liu X, Yang J, Wei M. In situ synthesis of $Na_2Ti_7O_{15}$ nanotubes on a Ti net substrate as a high performance anode for Na-ion batteries. Chem Commun. 2015;51:9298–9300.

[15] Miyauchi M, Tokudome H, Toda Y, Kamiya T, Hosono H. Electron field emission from TiO_2 nanotube arrays synthesized by hydrothermal reaction. Appl Phys Lett. 2006;89:043114.

[16] Ghicov A, Tsuchiya H, Hahn R, Macak J M, Muñoz A G. Schmuki P. TiO_2 nanotubes: H+ insertion and strong electrochromic effects. Electrochem Commun. 2006;8:528–532.

[17] Jämsen E, Furnes O, Engesæter L B, Konttinen Y T, Odgaard A, Stefánsdóttir A, Lidgren L. Prevention of deep infection in joint replacement surgery. Acta Orthopaedica. 2010;81:660–666.

[18] Rasouli M R, Restrepo C, Maltenfort M G, Purtill J J, Parvizi J. Risk factors for surgical site infection following total joint arthroplasty. J Bone Joint Surg Am. 2014;96:e158.

[19] Yin Z, Sakamoto Y, Yu J, Sun S, Terasaki O, Xu R. Microemulsion-based synthesis of titanium phosphate nanotubes via amine extraction system. J Am Chem Soc. 2004;126:8882–8883.

[20] Bortun A I, Khainakov S A, Bortun L N, Poojary D M, Rodriguez J, Garcia J R, Clearfield A. Synthesis and characterization of two novel fibrous titanium phosphates $Ti_2O(PO_4)_2{\cdot}2H_2O$. Chem Mater. 1997;9:1805–1811.

[21] Liu J, Wei X. Supercritical synthesis of layered elongated hexagonal titanium phosphate nanoplates. RSC Adv. 2015;5:7798–7802.

[22] Liu J, Wei X, Yu Y, Song J, Wang X, Li A, Liu X-W, Deng W-Q. Uniform core–shell titanium phosphate nanospheres with orderly open nanopores: a highly active Brønsted acid catalyst. Chem Commun. 2010;46:1670–1672.

[23] Jia K, Pan B, Zhang Q, Zhang W, Jiang P, Hong C, Pan B, Zhang Q. Adsorption of Pb^{2+}, Zn^{2+}, and Cd^{2+} from waters by amorphous titanium phosphate. J Colloid Interface Sci. 2008;318:160–166.

[24] Wang X, Yang X, Cai J, Miao T, Li L, Li G, Deng D, Jiang L, Wang C. Novel flower-like titanium phosphate microstructures and their application in lead ion removal from drinking water. J Mater Chem A. 2014;2:6718–6722.

[25] Korösi L, Papp S, Dékány I. A layered titanium phosphate $Ti_2O_3(H_2PO_4)_2{\cdot}2H_2O$ with rectangular morphology: synthesis, structure, and cysteamine intercalation. Chem Mater. 2010;22:4356–4363.

[26] María J A, Candela V-A, Pedro L, José L T. Improving the electrochemical performance of titanium phosphate-based electrodes in sodium batteries by lithium substitution. J Mater Chem A. 2013;1:13963–13969.

[27] Park K-H, Mondal S, Ghosh S, Das S, Bhaumik A. Enhanced efficiency in dye-sensitized solar cells based on mesoporous titanium phosphate photoanode. Microporous Mesoporous Mater. 2016;225:255–260.

[28] Lu J-S. Corrosion of titanium in phosphoric acid at 250°C. Trans Nonferrous Met Soc China. 2009;19:552–556.

[29] Park J-W, Jang J-H, Lee C S, Hanawa T. Osteoconductivity of hydrophilic microstructured titanium implants with phosphate ion chemistry. Acta Biomater. 2009;5:2311–2321.

[30] Yada M, Inoue Y, Sakamoto A, Torikai T, Watari T. Synthesis and controllable wettability of micro- and nanostructured titanium phosphate thin films formed on titanium plates. ACS Appl Mater Interfaces. 2014;6:7695–7704.

[31] Mühlebach J, Müller K, Schwarzenbach G. Peroxo complexes of titanium. Inorg Chem. 1970;11:2381–2390.

[32] Wu J-M. Low-temperature preparation of titania nanorods through direct oxidation of titanium with hydrogen peroxide. J Cryst Growth. 2004;269:347–355.

[33] Yan Y Y, Gao N, Barthlott W. Mimicking natural superhydrophobic surfaces and grasping the wetting process: a review on recent progress in preparing superhydrophobic surfaces. Adv Colloid Interface Sci. 2011;169:80–105.

[34] Inoue Y, Uota M, Torikai T, Watari T, Noda I, Hotokebuchi T, Yada M. Antibacterial properties of nanostructured silver titanate thin films formed on a titanium plate. J Biomed Mater Res Part A. 2010;92:1171–1180.

[35] Matsumoto Y, Ida S, Inoue T. Photodeposition of metal and metal oxide at the TiO_x nanosheet to observe the photocatalytic active site. J Phys Chem C. 2008;112:11614–11616.

[36] Chernousova S, Epple M. Silver as antibacterial agent: ion, nanoparticle, and metal. Angew Chem Int Ed. 2013;52:1636–1653.

Chapter 11

Control of Hierarchical Structure of Crystalline Nanofibers Based on the Cooperative Phenomena of Functional Molecular Group as the Target of Expression of New Physical Properties: Creation of Molecular Conductors and Enhancement of Thixotropic Ability

Manami Iizuka, Rie Yamato and Atsuhiro Fujimori

Additional information is available at the end of the chapter

http://dx.doi.org/10.5772/63701

Abstract

In this chapter, a study on two kinds of examples of functional nanofibers has been introduced. In the first section, the formation of nanofiber morphology at a mesoscopic scale and molecular level stacking of a tetrathiafulvalene (TTF) derivative with a chiral group were investigated by the one-dimensional growth method in interfacial molecular films. Monomolecular films of a TTF derivative with a chiral borneol group display a two-dimensional phase transition at the air/water interface. The formation of nanonetwork domains is attributed to the organized aggregation of the TTF derivatives, which is a result of strong intermolecular interactions. In the second section, the formation of nanofiber morphology at a mesoscopic scale and molecular level packing of an amphiphilic diamide derivative with two hydrocarbons were investigated by the in interfacial molecular films. It has found that the growth of this nanofiber morphology is encouraged by the application of the epitaxial growth in the spin-cast film with 1 wt% layered silicate having long hydrocarbons. As mention above, it is found that mesoscopic morphological formation of super-hierarchical structure of the nanofibers having a crystalline arrangement at Å level can be induced conductivity and the thixotropy at macroscopic level.

Keywords: crystalline nanofiber, TTF derivative, diamide derivative with two hydrocarbons, thixotropic ability, one-dimensional growth

INTECH
open science | open minds

1. Morphological transition of a conductive molecular organization with non-covalent from nanonetwork to nanofiber

1.1. Introduction

Although the synthesis of the Langmuir–Blodgett (LB) film was first reported by Blodgett in 1934 [1], this technique has attracted a lot of attention in recent years, following its application in the formation of two-dimensional lamellae in linear polymers [2]. While there is a long history of development of this technique beginning with the studies on the optical and electronic properties of the LB films by Khun et al. [3, 4], direct observation of the folding of polymer chains in LB films using atomic force microscopy was a significant milestone, which revealed the phenomenon of "morphogenesis at the interface" [5, 6]. For the several examples of "morphogenesis at the interface," the formation of nanofibers [7], nanowires [8, 9], nanospheres [10], nanocoils [11], nanoribbons [12], sea-island structures [13, 14], rods [15], gyroids [16], lamellae [17], and honeycombs [18] are noteworthy. Simultaneous control of the mesoscopic morphology (at the sub-micron level) of a non-covalent molecular organization, along with control of the molecular arrangement and packing structure (at the sub-nanometer level), is essential for the construction of next-generation quantum devices and medical materials. In other words, the molecular arrangement and morphology need to be controlled at different dimensional scales.

In this chapter, the formation of a nanofiber morphology at the mesoscopic scale and the molecular level stacking of a tetrathiafulvalene (TTF) derivative with a chiral group were investigated by the method of one-dimensional molecular film growth (**Figure 1(a)**). The fabrication of a nanometer-scale thin film having a fibril texture, which was achieved by incorporating a phenyl group in the TTF derivatives, has been reported previously [19]. However, the strong interactions between the molecules, which is a consequence of the competing forces of π–π interactions (of TTFs and phenyl rings) and hydrogen-bonding (of urethane bonding units), inhibited accurate morphological control. Therefore, in our previous study, we attempted to substitute the phenyl ring with the bulky and chiral borneol group [20], in order to minimize these strong intermolecular interactions. However, since TTF derivatives containing borneol groups are amorphous, top-down fabrication of thin films and control of the molecular arrangement were quite difficult.

Organized molecular films [21] have been developed as potential candidates for the synthesis of biomimetic models [22] and molecular electronic devices [23, 24], which are of considerable technological interest [25–29]. In addition to the interaction of lipids and proteins with hydrocarbons, various amphiphiles containing functional groups (including π- and d-electron systems), and polymerizable groups have been synthesized to obtain monolayer assemblies with well-defined molecular arrangements [30]. Therefore, this technology could potentially provide an effective means for attaining hierarchically precise and regular structures.

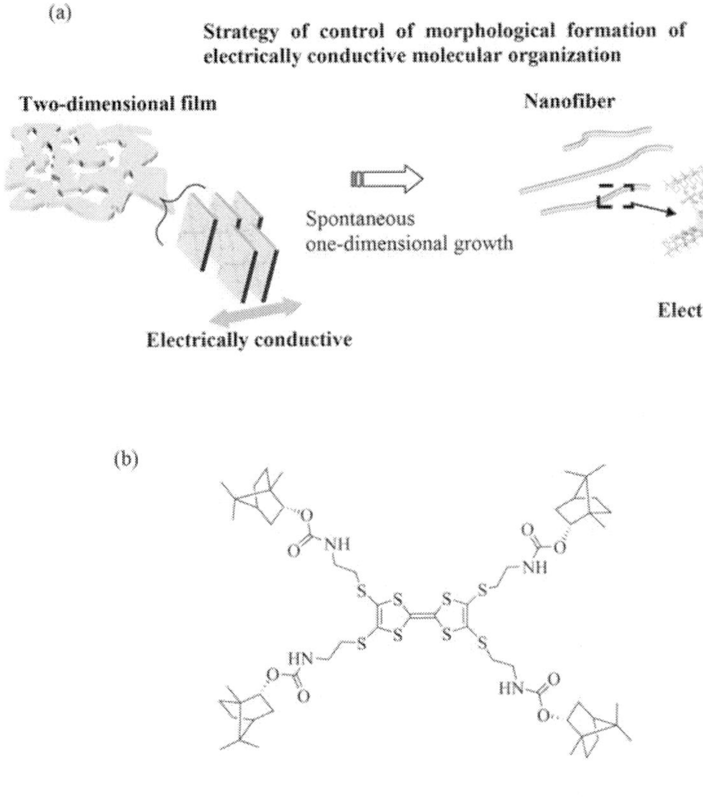

Figure 1. (a) Formation of nanofiber morphology at mesoscopic scale and molecular level stacking and (b) chemical structure of TTF-4Bor.

In the present chapter, a new, conducting, four-armed amphiphilic compound (a TTF derivative with chiral borneol groups (TTF-4Bor, **Figure 1(b)** [20]) was synthesized. The monolayer behavior, molecular arrangement, and surface morphology of organized molecular films of TTF-4Bor were investigated by analysis of the surface pressure–area (π–A) isotherms, in-plane and out-of-plane X-ray diffraction (XRD) profiles, and atomic force microscopy (AFM) images. Further, morphogenesis was encouraged by applying the one-dimensional growth method to an LB film of TTF-4Bor in distilled water, under low surface pressure conditions. The size of the nanofibers thus obtained was controlled by variation of the subphase temperature. Since the internal structure of the mesoscopic nanofiber is comprised of stacked TTF planes at a molecular level, effective electrical conduction is expected along the direction of stacking.

1.2. Experimental section

1.2.1. Synthesis of TTF-4Bor (Scheme S1)

A three-necked flask with a stirrer was purged with nitrogen, followed by the addition of L(-)-borneol (compound **1**, 20.5 g, 1.33 mmol), 2-chloroethyl isocyanate (compound **2**, 16.8 g, 159 mmol), and anhydrous tetrahydrofuran (THF, 300 mL). To this, 24 drops of dibutyltin dilaurate were added, and the mixture was stirred overnight in an oil bath at 50°C. The solvent was then evaporated under reduced pressure, and the residue was purified by column chromatography over silica gel (chloroform/ethyl acetate = 9:1). The product (compound **3**) was obtained as a white solid with a yield of 84% (29.07 g, 112 mmol).

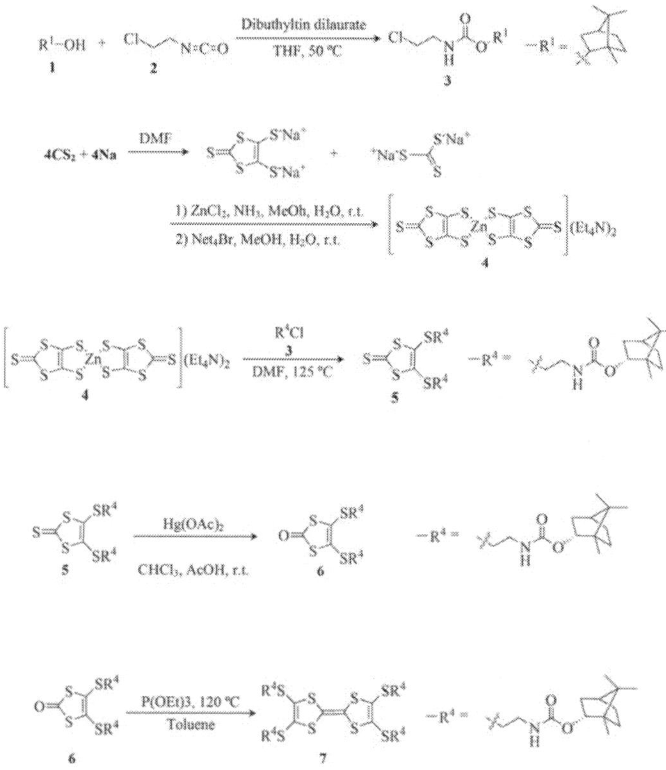

Scheme S1. Synthesis of TTF-4Bor.

Synthesis of Compound **4**: To a 500-mL four-necked flask (with a stirrer, purged with nitrogen), shredded sodium (23.0 g, 1.00 mol) and carbon disulfide (180 mL, 3 mol) were added and stirred. Then, the four-necked flask was placed in an ice bath, and *N,N*-dimethylformamide

(DMF) (200 mL) was added in drops over 4 h through a dropping funnel. The mixture was removed from the ice bath, brought to room temperature, and stirred overnight. This was again cooled in an ice bath, and methanol (50 mL) was added slowly through a dropping funnel. The contents of the four-necked flask were then transferred to a 3 L conical flask, and a mixture of methanol (200 mL) and deionized water (250 mL) was added in drops, followed by the gradual addition of a solution of zinc chloride (20 g, 0.15 mol) in concentrated aqueous ammonia (500 mL). To this, a solution of tetraethylammonium bromide (106 g, 0.50 mol) in deionized water (500 mL) was added in drops over a period of 4 h (in two portions), and the mixture was stirred overnight. This was suction-filtered through a Kiriyama funnel, and the residue was washed with water (500 mL), isopropanol (200 mL), and diethyl ether (200 mL). This residue was dissolved in acetone, and the resulting solution was evaporated under reduced pressure. Red crystals of the desired product (compound **4**) were obtained with 97.6% yield (87.7 g, 0.12 mol).

Synthesis of Compound **5**: In a 100-mL three-necked flask (with stirrer, purged with nitrogen), compound **4** (5.53 g, 7.70 mmol), compound **3** (10.00 g, 38.49 mmol), and DMF (90 mL) were added and stirred overnight in an oil bath at 125°C. Following concentration of the reaction mixture, the organic layer was extracted thrice with chloroform and distilled water, and dried using anhydrous magnesium sulfate. The magnesium sulfate was removed by filtration, and the filtrate was purified by column chromatography over silica gel (chloroform/ethyl acetate = 47:3). The desired product (compound **5**) was obtained as a yellow solid with 72.3% yield (7.19 g, 11.15 mmol).

Synthesis of Compound **6**: To a 100-mL three-necked flask (with a stirrer, purged with nitrogen), compound **5** (1.00 g, 1.55 mmol), chloroform (9.70 mL), acetic acid (3.23 mL), and mercury acetate (II) (1.72 g, 4.13 mmol) were added, and this mixture was stirred overnight at room temperature. The mixture was then subjected to Celite filtration, and the filtrate was extracted thrice using chloroform and distilled water. The chloroform layer was washed thrice with a saturated aqueous sodium bicarbonate solution and dried over anhydrous magnesium sulfate. This mixture was purified by column chromatography over silica gel (ethyl acetate). The desired product (compound **6**) was obtained as a yellow solid with a yield of 84.1% (0.82 g, 1.3 mmol).

Synthesis of Compound **7** *(TTF-4Bor)*: In a 50-mL three-necked flask (with a stirrer, purged with nitrogen), a mixture of compound **6** (0.82 g, 1.30 mmol) and triethyl phosphite (1.5 mL) was heated in an oil bath at 120°C with stirring for half day. The triethyl phosphite was removed under reduced pressure, and the remaining solid was purified by column chromatography over silica gel (hexane/ethyl acetate = 3:2). The desired final product (compound **7**, TTF-4Bor, **Figure 1(b)**) was obtained as a yellow solid with a yield of 63% (0.51 g, 0.41 mmol). The chemical structure of this compound has been determined by NMR and IR spectroscopy previously [20].

1.2.2. Formation of monolayers of TTF-4Bor on the water surface

Monolayers were formed by spreading a toluene solution of TTF-4Bor (~1.0×10^{-4} M) on the surface of distilled water (resistivity: approximately 18.2 MΩ cm). After evaporation of the toluene for 5 min, surface pressure–area (π–A) isotherms were recorded at compression speeds ranging from 0.8 to 4.8 $cm^2\ min^{-1}$. The air/water interface was kept at a constant temperature of 15°C by circulation of thermostated water around the trough. Measurement of the monolayer properties and LB film transfer were carried out in a USI-3-22 Teflon-coated LB trough (USI Instruments).

1.2.3. Study of the surface morphology and estimation of the molecular arrangement

The surface morphologies of the transferred films were observed using a scanning probe microscope (Atomic Force Microscopy, SII Nanotechnology, SPA300 with SPI-3800 probe station), and microfabricated rectangular Si cantilevers with integrated pyramidal tips, by applying a constant force of 1.4 N m^{-1}. In this chapter, AFM observations were carried out in the tapping mode. XRD samples were transferred onto a glass substrate by the LB method (20 layers, subphase temperature of 15°C, and surface pressures of 5 and 35 mN m^{-1}). The large spacing between the layers in the films was measured using an out-of-plane X-ray diffractometer (Rigaku, Rint-Ultima III, CuKα radiation, 40 kV, 30 mA) equipped with a graphite monochromator. The in-plane spacing of the two-dimensional lattice of the films was determined using an X-ray diffractometer with different geometrical arrangements [31, 32] (Bruker AXS, MXP-BX, CuKα radiation, 40 kV, 40 mA, a customized instrument) and equipped with a parabolic-graded multilayer mirror. The X-rays were incident at an angle of 0.2°, and the films were scanned at a speed of 0.05°/80 s, as a result of which the in-plane XRD measurements had monomolecular resolution.

1.3. Results and discussion

1.3.1. Monolayer behavior and surface morphology of TTF-4Bor

Figure 2 shows the π–A isotherms of a TTF-4Bor monolayer at the air/water interface at 15°C, along with the estimated collapse surface pressure and the limiting area. TTF-4Bor forms an expanded phase at low surface pressures and an extremely condensed phase at high surface pressures. These results indicate that the TTF-4Bor film undergoes a two-dimensional transition, from parallel to normal orientation on the water surface.

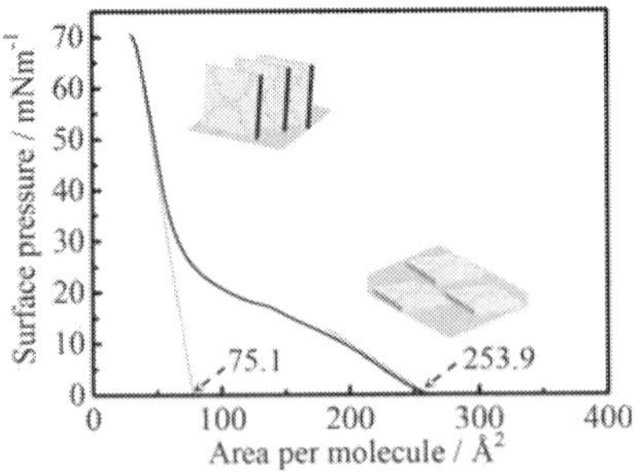

Figure 2. Surface pressure–area isotherm at the air/water interface for the TTF-4Bor monolayer.

http://dx.doi.org/10.5772/63701

Figure 3 shows the AFM images of LB monolayers of TTF-4Bor (Z-type) on mica, transferred at 35 and 10 mN m^{-1}. At low surface pressures, sparsely dotted nanodomains with a height of about 1 nm are observed, confirming the flat-on orientation of the molecular planes. In the high pressure regions, however, submicron network domains are formed, and the height information indicates that the conformation of the TTF derivatives is normal to the plane of the film. It has been suggested that the mesoscopic morphogenesis is attained by aggregation, which is based on the competition between the π–π interaction of the TTF molecular planes, and the strong interaction of hydrogen bonds between the urethane bonding sites. This hypothesis is supported by the infrared (IR) spectra (**Figure 3(c)**), which exhibit a shift of the band attributed to the carbonyl group toward lower wavenumbers, indicating the formation of hydrogen bonds [19, 20].

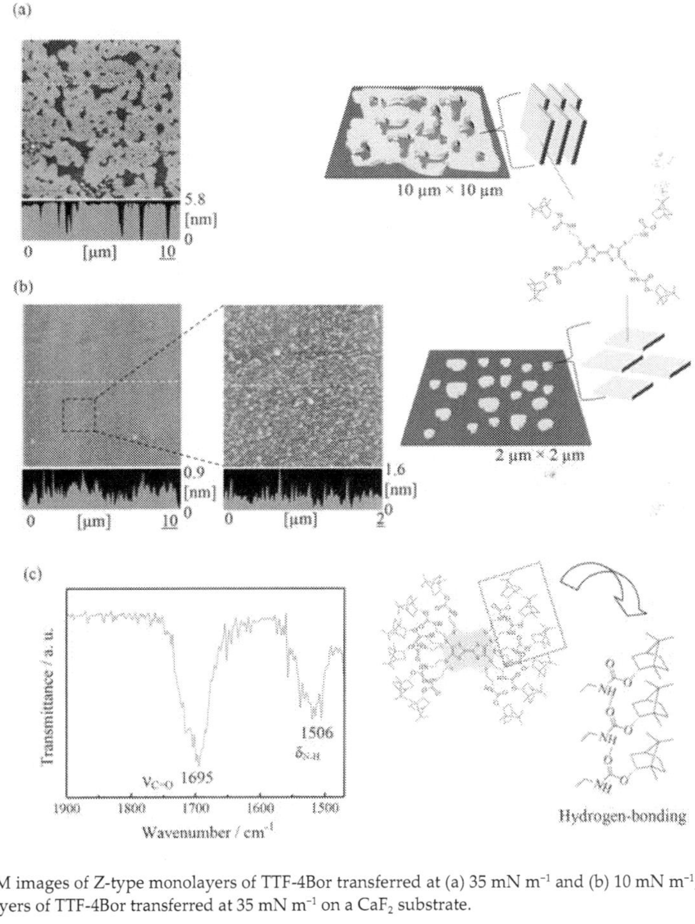

Figure 3. AFM images of Z-type monolayers of TTF-4Bor transferred at (a) 35 mN m^{-1} and (b) 10 mN m^{-1}; (c) IR spectra of LB multilayers of TTF-4Bor transferred at 35 mN m^{-1} on a CaF_2 substrate.

1.3.2. Molecular arrangement in organized molecular films of TTF-4Bor

In order to estimate the crystallinity and periodicity of the molecules in multilayers of TTF-4Bor, in-plane and out-of-plane XRD analyses of LB multilayers were carried out. In-plane XRD profiles of multilayers transferred at 35 and 10 mN m^{-1} are shown in **Figure 4**. This technique provides information regarding molecular arrangement at a sub-nanometer scale, and the internal fine structure of the mesoscopic morphology. At low surface pressures, a clear periodic structure was not observed, indicating that the film might be amorphous. On the other hand, in the LB multilayers formed at high surface pressures, a clear periodic structure with regular molecular packing was confirmed. The short spacing value of 4.1 Å appears to correspond to the stacking of the TTF molecular planes based on π–π interactions. A similar in-plane packing system based on π–π interactions, established by in-plane XRD, was reported in conducting organized molecular films of metal $(dmit)_2$ charge transfer complexes [33]. Dense stacking of molecular planes has been known to induce electrical conductivity along the direction of stacking.

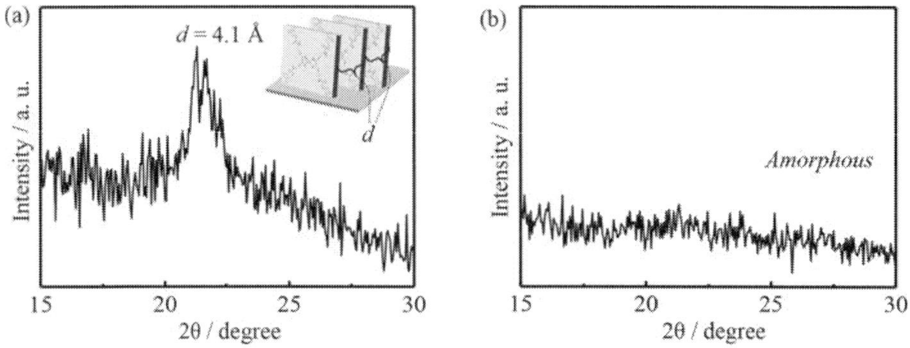

Figure 4. In-plane XRD profiles of LB multilayers (20 layers) of TTF-4Bor transferred at (a) 35 mN m^{-1} and (b) 10 mN m^{-1}.

Figure 5 shows the out-of-plane XRD profiles for LB multilayers (20 layers) of TTF-4Bor, transferred at 35 and 10 mN m^{-1}. In the multilayers fabricated under high surface pressure conditions, a d_{001} layer corresponding to the end-on orientation of TTF-4Bor was observed along the *c*-axis. The formation of the highly regular layered structure is believed to have occurred during the transfer of the interfacial monomolecular film to a solid, by the LB method. At low surface pressures, a layer spacing of ~5 nm is observed in the multilayers, which exhibit molecular organization along with disorder in the *ab*-plane, and regularity along the *c*-axis. Therefore, rearrangement of the molecular groups is expected to take place during multilayer formation, along with a change in the conformation from flat-on to end-on, resulting in layered regularity in the *c*-direction. These results are summarized in **Figure 6**.

http://dx.doi.org/10.5772/63701

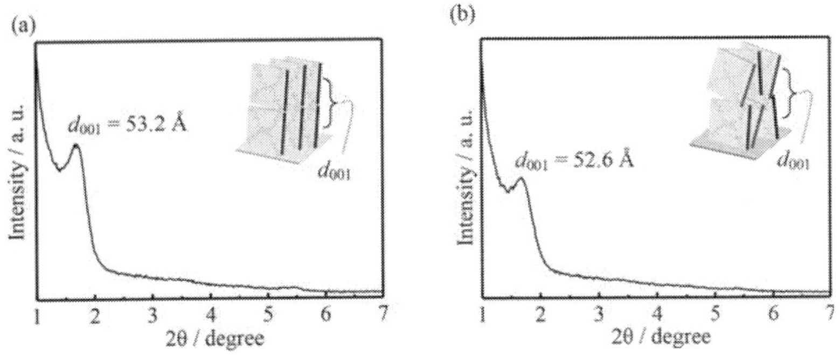

Figure 5. Out-of-plane XRD profiles of LB multilayers (20 layers) of TTF-4Bor transferred at (a) 35 mN m^{-1} and (b) 10 mN m^{-1}.

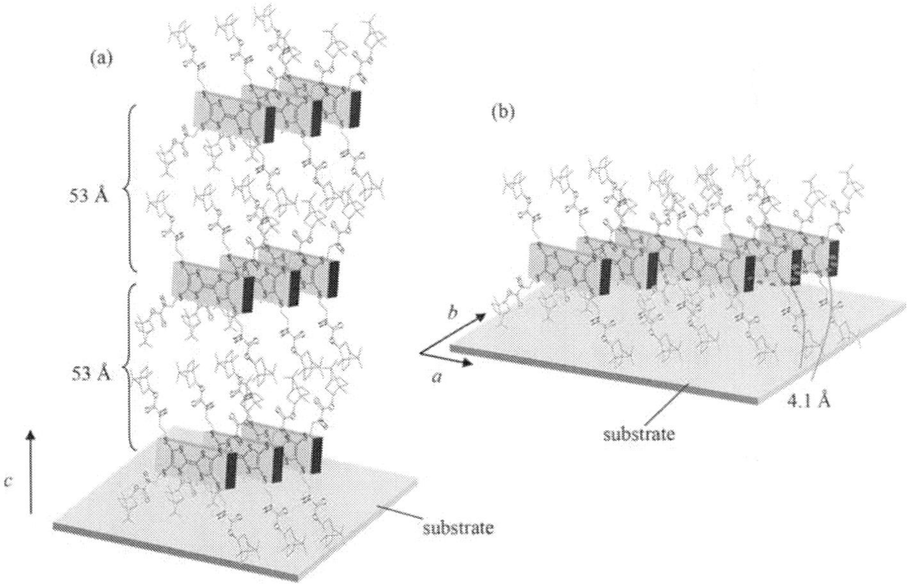

Figure 6. Schematic illustration of the (a) molecular arrangement of TTF-4Bor along the *c*-axis and (b) molecular stacking model of LB multilayers of TTF-4Bor in the *ab*-plane.

1.3.3. *Morphological changes from a two-dimensional film to nanofibers of TTF-4Bor*

At present, there are limitations in the ability to achieve one-dimensional growth of a nanofiber consisting of tightly stacked electrically conducting molecules. Although stacked conducting molecular planes are obtained at high surface pressures, morphologies at the mesoscopic scale are too developed due to the strong forces of aggregation, with the presence of long-range order between molecules, as a result of the competitive effect of π–π interactions and hydrogen bonding. Hence, although the molecules cannot be arranged under low surface pressure conditions, the disordered molecular groups can be rearranged by multilayer formation, owing to their strong aggregation tendency. In this chapter, the technique of one-dimensional growth at the air/water interface was adopted, which involved transfer of the film at low surface pressures with reduction of the compression speed (by a factor of 8), inducing a spontaneous growth structure at the interface. **Figure 7** shows a comparison of the π–A isotherms, at different compression speeds (4.8 and 0.6 cm^2 min^{-1}), and a schematic illustration of molecular aggregation. The tendency to expand at low pressures and condense at higher pressures is conspicuous in the π–A isotherms measured at a low compression rate (0.6 cm^2 min^{-1}).

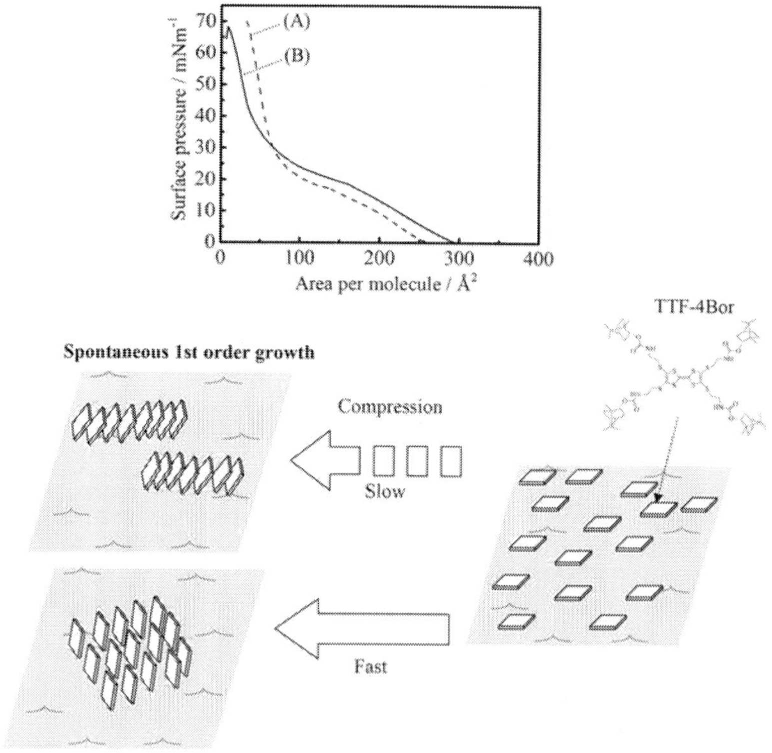

Figure 7. Surface pressure–area isotherms at different compression speeds (4.8 cm^2 min^{-1} [dashed line] and 0.6 cm^2 min^{-1} [solid line]) of monolayer of TTF-4Bor at the air/water interface.

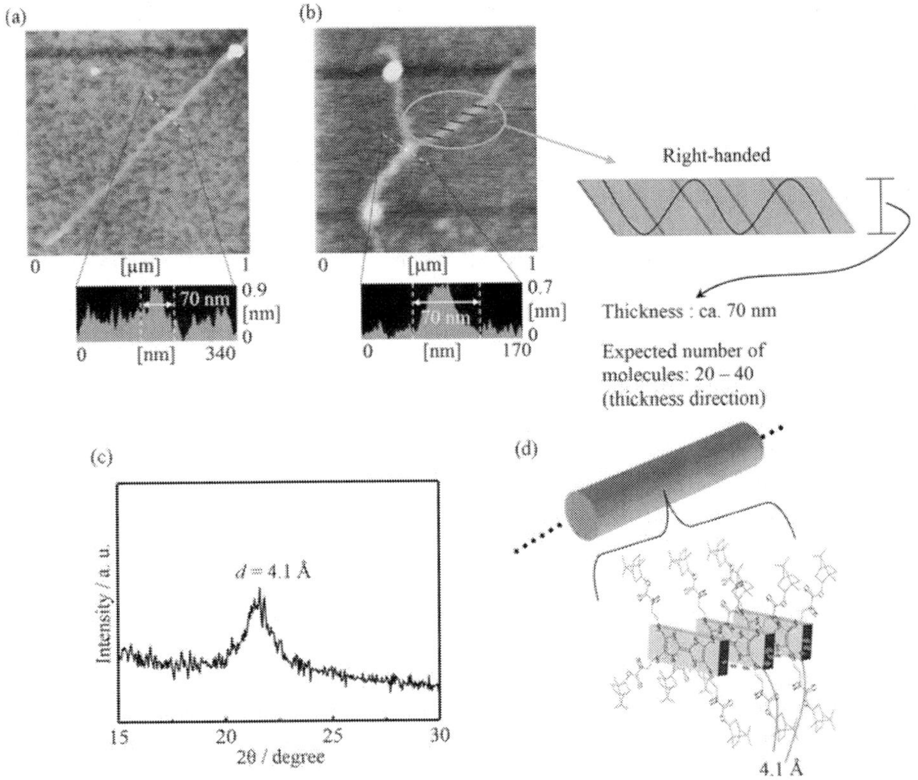

Figure 8. AFM images of Z-type monolayers of (a) nanofiber and (b) helical nanofiber forms of the one-dimensional growth TTF-4Bor. (c) In-plane XRD profiles of LB multilayers (20 layers) of TTF-4Bor (at 10 mN m^{-1} and compression speed 0.6 cm^2 min^{-1}), and (d) a schematic model for their molecular packing.

In **Figure 8**, AFM images of the Z-type monolayers of TTF-4Bor after one-dimensional growth are shown, along with the in-plane XRD profiles of multilayers of TTF-4Bor transferred at 10 mN m^{-1} with a compression speed of 0.6 cm^2 min^{-1}. In this system, there is a clear transition from a mesoscopic morphology to a nanofiber shape (thickness ~70 nm). While many of the nanofibers are linear (**Figure 8(a)**), some fibers exhibit a branching morphology (**Figure 8(b)**) and some others grew as a right-handed spiral. This morphological formation is expected to have occurred to minimize the steric hindrance from the four functional groups in the TTF derivative. This material exhibited circular dichroism (CD) for every adsorption band [20], and the Cotton effect centered around 380 and 530 nm was attributed to the helical dipole coupling between TTF rings through the formation of a network of hydrogen bonds [34–38]. In-plane

XRD reveals multilayers of these integrated nanofiber films, with the formation of stacked conducting molecular planes with a spacing of 4.1 Å (**Figures 8(c)** and **(d)**). It is expected that the promotion of spiral morphogenesis based on steric hindrance of chiral functional groups is dependent on the enhancement of the molecular mobility at the air/water interface. Therefore, the formation of nanofibers was studied at different subphase temperatures (**Figure 9**). It is seen that an increase in the subphase temperature results in an increase in the diameter of the fibers, rather than the formation of the helical structure, which could be due to an increase in the aggregation force between the molecules. The diameters of the nanofibers formed at 15, 20, and 25°C, are 70, 125, and 190 nm, respectively. From the π–A curves, these correspond to aggregates of 20–40 molecules at 15°C, 30–60 molecules at 20°C, and 50–90 molecules at 25°C. The nanofiber morphology was retained upon inclusion of 2,3,5,6-tetrafluoro-7,7,8,8-tetracyanoquinodimethane (F_4-TCNQ) acceptor molecules to this system at a molecular ratio of 1:1 (**Figure 10**). Since F_4-TCNQ is not amphiphilic, mixing it with the TTF derivative results in the formation of a charge transfer (CT) complex and interfacial nanofiber films. The retention of the nanofiber morphology after the inclusion of F_4-TCNQ is because the structural formation is related to the stacking arrangement of the TTF molecular planes, and first-order spontaneous growth of mesoscopic nanofibers. The CT complex exhibited a broad absorption band in the IR region (2500–3500 cm^{-1}), which was attributed to a CT between the electron-donating TTF and the electron-accepting F_4-TCNQ moieties [39–41]. **Figure 10** shows the color change of the solution to support the CT complex formation, and further, the AFM image and in-plane XRD profile also show evidences of maintaining of a mesoscopic fiber form and stacking of a conductive molecular plane. In addition, **Figure 10(c)** also shows UV spectrum of charge transfer complex of TTF-4Bor:F4-TCNQ = 1:1 in solution. Absorption band near 300 nm corresponds to the TTF-4Bor, and the band of 400 nm except the shoulder peak corresponding to F4-TCNQ. Absorption band at 700–800 nm corresponds to the anionic radical of the F4-TCNQ, and shoulder-shaped absorption band around 400 nm corresponds to the cationic radical of TTF [41]. From the above, the formation of a charge-transfer complex in solution is supported.

As detailed in this chapter, the amphiphilic TTF derivative with chiral borneol groups are closely packed in the nanofiber morphology at the air/water interface. Although the densely packed molecular arrangement is easily attained by standard film fabrication methods at high surface pressures, mesoscopic nanofiber morphology is obtained by spontaneous one-dimensional growth at the air/water interface (**Figure 11**). The dense stacking structure of these molecular planes induces electrical conduction, and nanofibrous morphogenesis can be deployed in quantum devices, medical applications, etc. It can be seen that the control of the mesoscopic morphology at the air/water interface provides several variations and possibilities in the chapter of molecular organization, such as helical morphogenesis and size control of nanofibers [42].

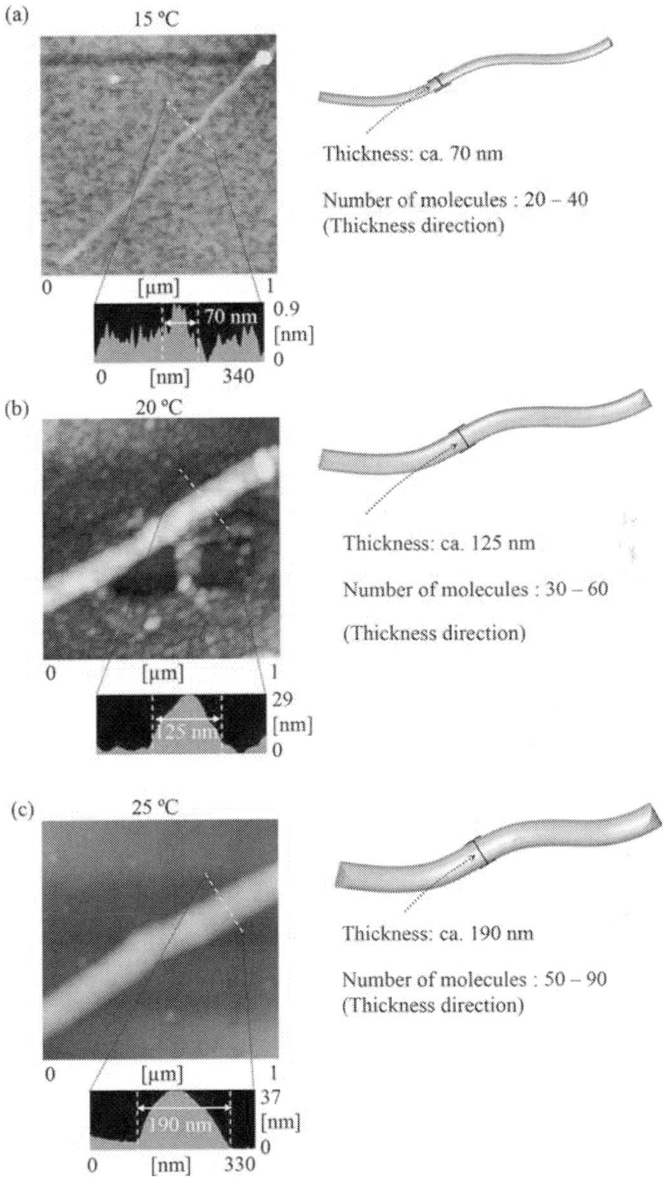

Figure 9. AFM images of Z-type monolayers of the one-dimensional growth of TTF-4Bor transferred at (a) 15°C, (b) 20°C, and (c) 25°C.

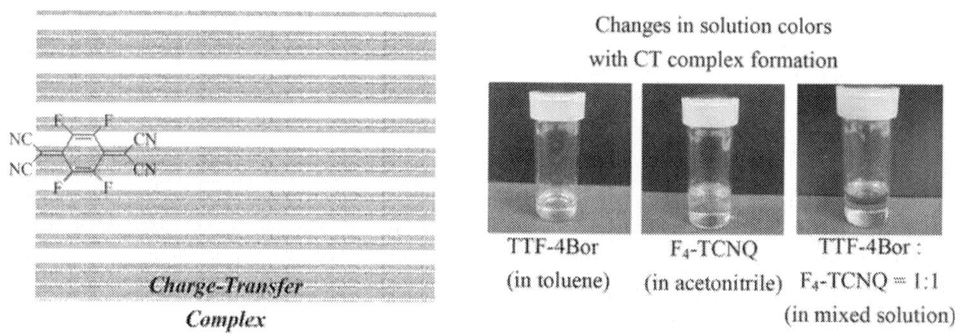

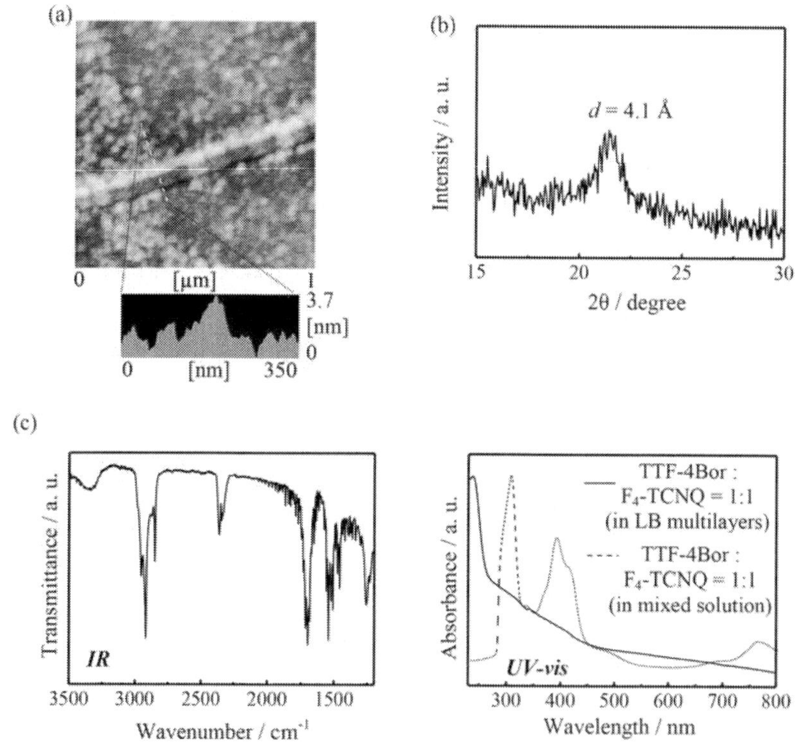

Figure 10. (a) AFM image of Z-type monolayer, (b) in-plane XRD, and (c) IR and UV–Vis spectra of the LB multilayers (20 layers) of TTF derivative/F_4-TCNQ = 1:1 charge transfer complex.

http://dx.doi.org/10.5772/63701

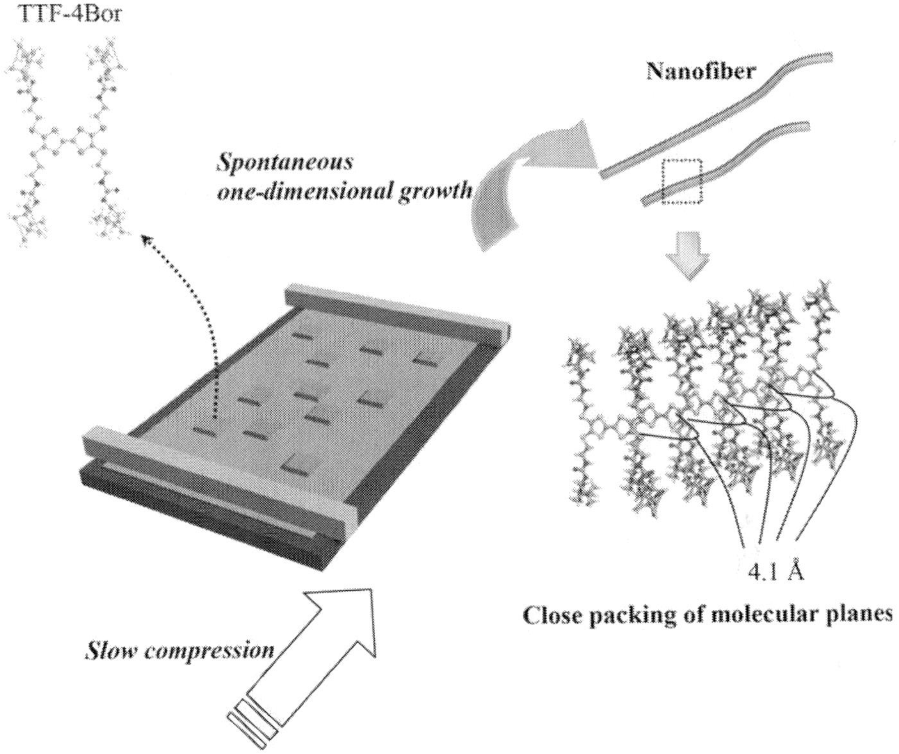

Figure 11. Schematic models for the formation of spiral morphology at the mesoscopic scale, and molecular level stacking of TTF-4Bor by the one-dimensional growth method.

1.4. Conclusions

In this chapter, monolayer behavior on the surface of water, mesoscopic morphological formation, and molecular arrangement in LB multilayers of the conductive TTF derivative with chiral groups were investigated. For this purpose, π–A isotherms, in-plane and out-of-plane XRD, and AFM measurements were carried out. From the AFM analysis, the formation of submicron networks and nanofiber morphology of the conducting molecular organization was confirmed at different monolayer compression rates. The in-plane and out-of-plane XRD measurements elucidated the formation of highly ordered layered structures and close-stacking of molecular planes due to π–π interactions.

Nanofibers of the TTF derivative were obtained by applying the one-dimensional growth method at low compression speeds and low surface pressure conditions. Under these conditions, the formation of helical nanofibers was also observed. These mesoscopic helical nanofibers, whose internal structure consists of stacked conducting planes, are expected to display spontaneous electrical conduction by electromagnetic induction. Hence, these materials are likely candidates for future innovation in the molecular device and healthcare industries.

2. Morphological control of crystalline nanofiber derived from amphiphilic diamide derivative which induces the thixotropic ability

2.1. Introduction

In the molecular and materials chemistry, the concept of "hierarchy [43]" is very important. In usual, the molecular level orientation and arrangements beyond the level of mesoscopic morphogenesis cannot affect the level of the macroscopic physical properties. However, there are cases that molecular groups with crystalline array achieved a fibrous growth by the cooperative phenomenon of molecules, and its network form occurs thixotropic ability [43]. If this crystalline fiber network contacts to other medium and induces the increase of that viscosity, the corresponding molecule is acting as a thixotropic agent beyond the size hierarchy of materials.

In addition, current nanotechnology in the chemistry and biophysics fields has undergone a remarkable development through the "discovery of the nanosized architecture" such as carbon nanotubes [44], "their deployment to the nanocomposite materials [45–48]", and "re-attention to the bottom-up technologies as the classical LB method [1]". Performance enhancement of analytical techniques, such as scanning probe microscopy and X-ray diffraction, has also helped strongly to the development of nanoscience.

By the way, the thixotropic phenomenon, which solidification and the flow are, respectively, occurred by standing and application of an external forces, is observed in familiar mayonnaise, ketchup, etc. Commercially available anti-sedimentation agent and liquid dripping inhibitors are also used as thixotropic agents. In addition, an additive which can be solidified the cooking waste oil and discarded it as a solid is also corresponding to a thixotropic agent. Traditionally, the main raw material of the thixotropic agent was 12-hydroxystearic acid. In this case, an occurrence origin of thixotropic properties is the fiber growth by 12-hydroxystearic acid molecule at micrometer level. The increase in viscosity has occurred by the contact of this developed microfiber to the corresponding medium. In other words, relationship between fiber growth and occurrence of thixotropic properties is almost equal. However, although the 12-hydroxy stearic acid can form a fibrous morphology in the view of molecular science, stearic acid does not form a fibrous form. It is well-known that stearic acid is the standard material of monolayer on the water surface, and corresponding LB film. Although stearic acid is a crystalline compound, its mesoscopic form is a sheet-like morphology. In other words, the influence of the hydrogen bonding between hydroxyl groups at the 12-position to the formation of the fibrous morphology is remarkable. However, the hydrogen bonding itself is also interaction at the molecular level. It cannot be asserted to affect the mesoscopic morphogenesis and macroscopic thixotropic by only the presence of a simple hydrogen bonding. However, hydrogen bonding "between biological polymers that forms the human body" and "between the polyamide fibers which is raw materials of a garment," reality induces the macroscopic physical properties. In order to attain an understanding of these phenomena, the material evaluation by the hierarchical point of view will be essential by making full use of nanotechnological analysis developed in recent years. In addition, discussion ability that be considered in connecting these phenomena will be extremely important. However, discussion to the 12-hydroxystearic acid itself has already been a thing of the past on the development of academic field of modern chemistry.

In the present chapter, amphiphilic diamide derivative which obtained by condensation reaction of 12-hydroxystearic acids and hexamethylenediamine was synthesized. The monolayer behavior, molecular arrangement, and surface morphology of organized molecular films of diamide derivative with two hydrocarbons were investigated by analysis of the π–A isotherms, in-plane and out-of-plane XRD profiles, and AFM images. Further, morphogenesis was encouraged by applying the epitaxial growth to a spin-cast film of diamide derivative. The form of the nanofibers thus obtained was indicated the linearly developed shape. Since the internal structure of the mesoscopic nanofiber is comprised of packed hydrocarbons at a molecular level, effective molecular packing is expected along the fiber growing direction.

2.2. Experimental section

2.2.1. Synthesis and characterization of amphiphilic diamide derivative with two hydrocarbons

In order to obtain the material used in this chapter, a condensation reaction of 12-hydroxystearic acid and hexamethylenediamine (2 mol: 1 mol) was performed. The obtained material was purified by recrystallization, and a removal of impurity checked by thermal analysis. **Figure 12(a)** shows thermogravimetric (TG) curves of this material under the N_2 purge and in the air atmosphere. Under the N_2 purge and in the air atmosphere, the decreasing of weight of material is started at 250 and 350°C, respectively. Further, there is a residual of heated material until 500°C and in the air atmosphere. That is to say, this material decomposes like a certain polymer at 250°C and do not sublimate as substance with low molecular weight. In other words, it finds that this compound is an extreme high heat resistance by the influence of hydrogen bonding. **Figure 12(b)** shows differential scanning calorimetric (DSC) thermogram of diamide derivative with two hydrocarbons. Only one transition peaks in both heating and cooling processes exist at around 150 and 135°C, respectively. In the case of unpurified sample, transition peaks of reacted substances in heating process are indicated at around 77 and 42°C. Therefore, it was considered that unreacted 12-hydroxystearic acid and hexamethylenediamine have been removed by recrystallization purification.

2.2.2. Characterization of bulk and their spin-cast film

The spin-cast films of diamide derivative with two hydrocarbons and the mixture with organo-modified layered silicate were formed from xylene– ethanol (9/1, v/v) mixed solution over 100°C. Organo-modified montmorillonite (MMT) was fabricated by surface modification method with natural MMT and long-chain quaternary ammonium cation at oil/water interface [49]. Powder X-ray diffraction (XRD) measurement to the both bulk and cast film samples is

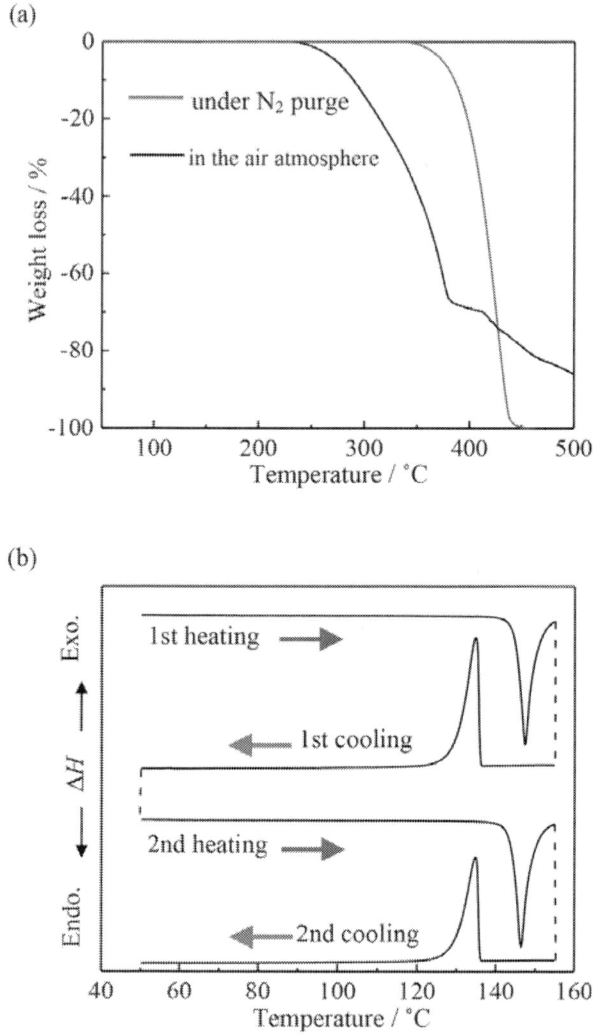

Figure 12. (a) TG curves of diamide derivatives having two hydrocarbons used in this study under the N_2 purge and in the air atmosphere. (b) DSC thermogram of purified diamide derivatives having two hydrocarbons (scanning rate; 10°C min^{-1}).

performed by X-ray diffractometer (Rigaku, Rint-Ultima III, CuKα radiation, 40 kV, 30 mA) equipped with a graphite monochromator. Infrared (IR) spectra of the sample are measured by IR spectrometer (Bruker AXS TENSOR II).

2.2.3. *Formation of monolayers of diamide derivative with two hydrocarbons on the water surface*

Monolayers were formed by spreading from a $CHCl_3$ solution including a small amount of trifluoroacetic acid (TFA) of diamide derivate having two-stearic chains with –OH groups ($\sim 1.0 \times 10^{-4}$ M) on the surface of distilled water (resistivity: approximately 18.2 MΩ cm). After evaporation of the $CHCl_3$ for 5 min, surface pressure-area (π–A) isotherms were recorded at compression speeds ranging from 4.8 $cm^2 min^{-1}$. The air/water interface was kept at a constant temperature of 3.5, 12, and 15°C by circulation of thermostated water around the trough. Measurement of the monolayer properties and LB film transfer were carried out in a USI-3-22 Teflon-coated LB trough (USI Instruments). Further, mixed monolayers on the aqueous buffer solution including Na^+ ion of diamide derivative and organo-MMT have been formed by co-spreading method of $CHCl_3$ solution with small amount of TFA and toluene solution, respectively.

2.2.4. *Study of the surface morphology and estimation of the molecular arrangement*

The surface morphologies of the transferred films were observed using a scanning probe microscope (Atomic Force Microscopy, SII Nanotechnology, SPA300 with SPI-3800 probe station), and microfabricated rectangular Si cantilevers with integrated pyramidal tips, by applying a constant force of 1.4 N m^{-1}. In this chapter, AFM observations were carried out in the tapping mode. XRD samples were transferred onto a glass substrate by the LB method (20 layers, subphase temperature of 15°C, and surface pressures of 35 mN m^{-1}). The large spacing between the layers in the films was measured using an out-of-plane X-ray diffractometer (Rigaku, Rint-Ultima III, CuKα radiation, 40 kV, 30 mA) equipped with a graphite monochromator. The in-plane spacing of the two-dimensional lattice of the films was determined using an X-ray diffractometer with different geometrical arrangements [31, 32] (Bruker AXS, MXP-BX, CuKα radiation, 40 kV, 40 mA, a customized instrument) and equipped with a parabolic graded multilayer mirror. The X-rays were incident at an angle of 0.2°, and the films were scanned at a speed of 0.05°/20 s, as a result of which the in-plane XRD measurements had monomolecular resolution.

2.3. Results and discussion

2.3.1. *Molecular arrangement and packing of diamide derivative including two-stearic chains with –OH groups*

Figures 13(a) and **(b)** shows the powder XRD profiles and IR spectrum of diamide derivative with two hydrocarbons in bulk. In powder XRD profile, the developed layer spacing is indicated in low angle side. In this case, the long spacing peaks until fifth-order reflection are clearly confirmed. Furthermore, bands of stretching vibration of N–H and O–H are shifted to relative low angle side in IR spectra by the influence on the hydrogen bonding. In addition, amide I and II bands are clearly confirmed at around 1640 and 1550 cm^{-1}.

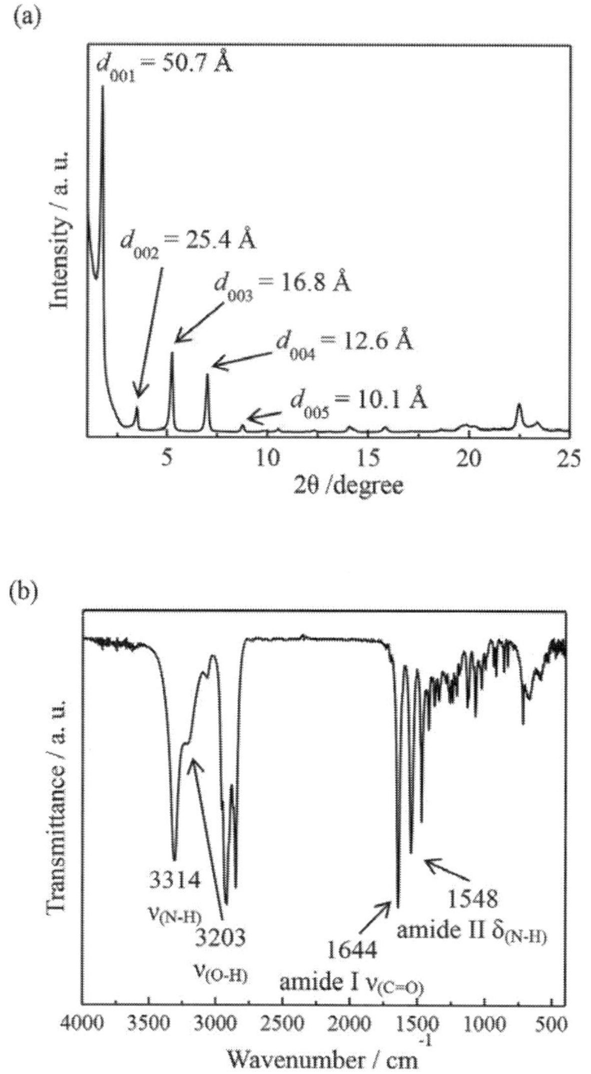

Figure 13. (a) Powder X-ray diffraction profile of diamide derivatives having two hydrocarbons in bulk. (b) IR spectra of diamide derivatives having two hydrocarbons in bulk.

Figure 14 shows the schematic illustration of two kind of possibility of molecular arrangement of diamide derivative with two hydrocarbons in bulk. In the case of model in **Figure 14(a)**, molecules form the extended chain conformation. On the other hand, bilayer structure is formed like a general surfactant molecules in **Figure 14(b)**. In both case, it finds that intermo-

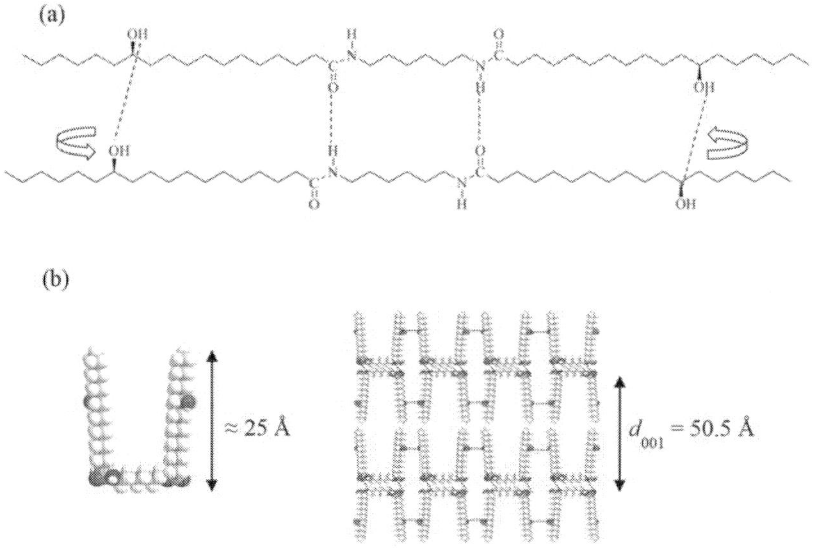

Figure 14. Schematic illustration of two kind of possibility of molecular arrangement of diamide derivatives having two hydrocarbons in bulk. (a) An extended chain model. (b) Double layered structure model.

lecular hydrogen bonding between hydroxyl groups and amide groups have formed by the results of IR. There are two kinds of possibility of molecular orientation that first-order reflection corresponds to tilted single-layer spacing or bilayer spacing. In this stage, both possibilities cannot be rejected. However, authors support bilayer conformation as **Figure 14(b)** according to the nature of this type of amphiphilic materials.

2.3.2. Monolayer behavior and surface morphology of diamide derivative with two hydrocarbons

Figure 15 shows the π–A isotherms of monolayer on the water surface of a diamide derivative with two hydrocarbons at 3.5, 12, and 15°C. At 15°C, isotherm of diamide derivative with two hydrocarbons indicates the overshoot shoulder and plateau region near the 10 mN m^{-1}. On the other hand, the number of two-dimensional phase transition increases in isotherms below 12°C. The origin of these two-dimensional phase transition can be inferred by the AFM measurement to monolayers which are transferred on the solid substrate. **Figure 16** shows the AFM images of LB monolayers of diamide derivative with two hydrocarbons (Z-type) on mica, transferred at 7, 10 and 35 mN m^{-1}. At low surface pressures region after first transition, continuously dotted nanodomains are observed. Next, thin fibrous morphology is confirmed after the second transition. Finally, in the high pressure regions, the developed fiber is formed as their monolayer feature. Therefore, it finds that an each two-dimensional transitions are corresponds to morphological changes of monolayer of diamide derivative with two hydrocarbons.

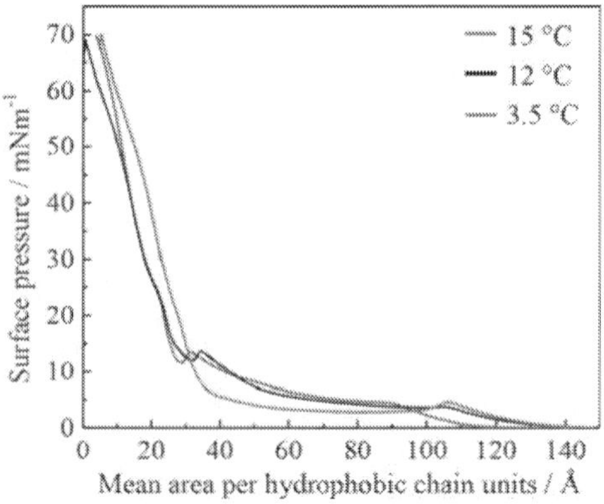

Figure 15. Surface pressure–area isotherms of monolayer on the water surface of diamide derivatives having two-hydrocarbons at 3.5, 12, and 15°C subphase temperature.

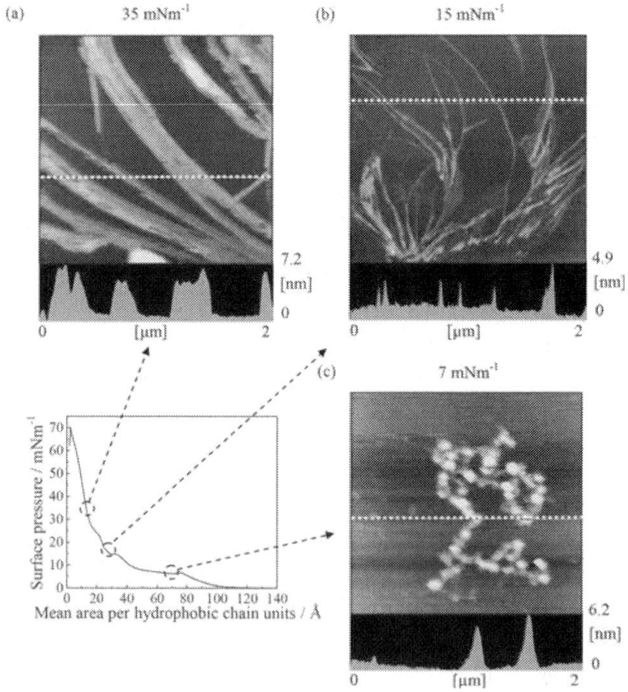

Figure 16. AFM images of monolayer on solid substrate of two-chain-type diamide derivatives transferred at (a) 35 mN m^{-1}, (b) 15 mN m^{-1}, and (c) 7 mN m^{-1} with corresponding isotherm at 12°C.

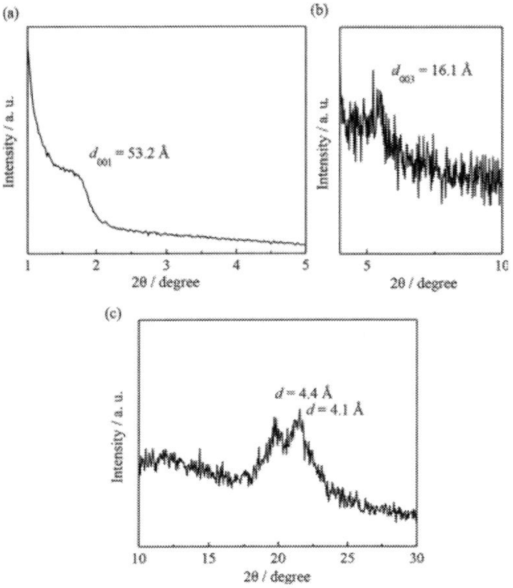

Figure 17. (a, b) Out-of-plane and (c) in-plane XRD profiles of LB multilayers of two-chain-type diamide derivatives transferred at 35 mN m^{-1} and 12°C.

2.3.3. Molecular arrangement in organized molecular films of diamide derivative having two hydrocarbons

In order to estimate the crystallinity and periodicity of the molecules in multilayers of diamide derivative having two hydrocarbons, out-of-plane and in-plane XRD analyses of LB multilayers were carried out. **Figures 17(a)** and **(b)** shows the out-of-plane XRD profiles for LB multilayers (20 layers) which are transferred at 35 mN m^{-1}. In the multilayers fabricated under high surface pressure conditions, it seems that a layer spacing along the *c*-axis at d_{001} = 53.2 Å is corresponding to the bilayer spacing. The shape of first-order reflection is unclear by the influence on the direct beam. However, third-order reflection has also been confirmed. It is expected that highly order layer structure has formed. In-plane XRD profiles of multilayers transferred at 35 mN m^{1} are shown in **Figure 17(c)**. This technique provides information regarding molecular arrangement at a sub-nanometer scale, and the internal fine structure of the mesoscopic fiber morphology. In the LB multilayers formed at crystalline phase, a clear periodic structure with regular molecular packing was confirmed. The short spacing values of 4.4 and 4.1 Å appear to correspond to the packing of the long hydrocarbon chains based on van der Waals interactions. A similar in-plane packing system based on van der Waals interactions between chains, established by in-plane XRD, was reported in organized molecular films of long-chain fatty acid [31]. The packing of hydrocarbons are assigned to two-dimensional orthorhombic system. These results are summarized in **Figure 18**. Due to the nature of the wide angle X-ray diffraction, the exact value of the first-order period (bilayer

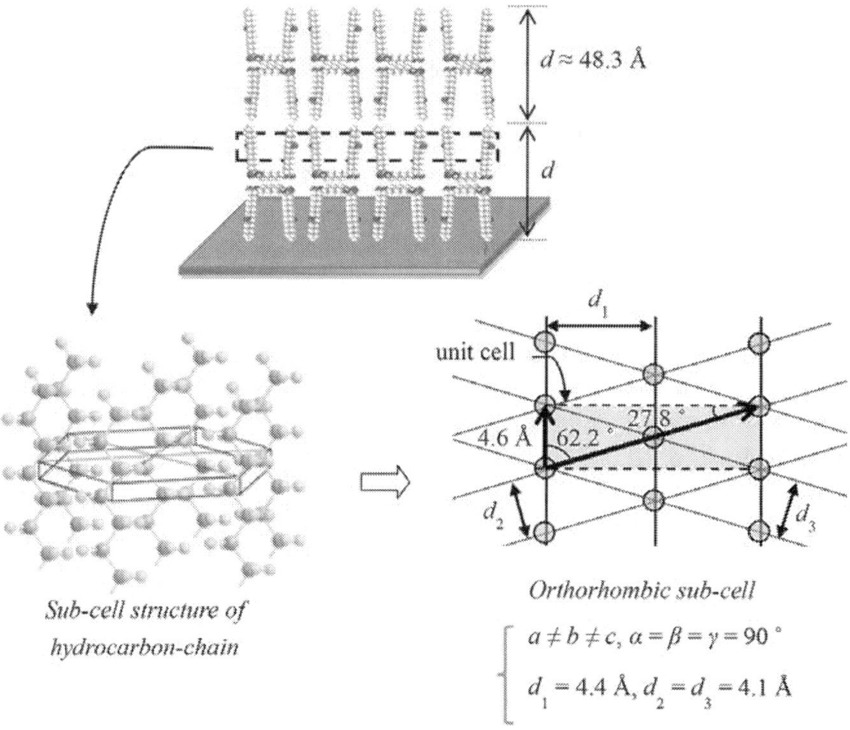

Figure 18. Schematic models of layer structure and sub-cell of hydrocarbons in LB multilayers of two-chain-type diamide derivatives.

spacing) corresponds to three times of $d_{0\,0\,3}$ value (48.3 Å) calculated from the third-order reflection, rather than $d_{0\,0\,1}$ value calculated from the first-order reflection. It finds that formation of the developed layer structure and sub-cell is the feature of this crystalline LB film. Further, bilayer conformation in this LB film is similar to the expected model of this molecule in the bulk state.

2.3.4. Liner morphological growth of nanofibers in a spin-cast film of diamide derivative having two hydrocarbons from layered silicate

At present, there are limitations in the ability to achieve one-dimensional growth of a nanofiber consisting of tightly packed molecules. Although the packed molecules are obtained in the two-dimensional interfacial film at high surface pressures, morphologies at the mesoscopic scale are too entangled and undulated due to the strong forces of aggregation, with the presence of long-range order between molecules, as a result of the competitive effect of van der Waals interactions and hydrogen bonding. Hence, if formation of the packed molecular arrangement is accelerated by the external forces at molecular level, the mesoscopic fiber

http://dx.doi.org/10.5772/63701

growth might be linearly and hierarchically developed. In this chapter, the technique of epitaxial growth from harmless layered material in the mesoscopic spin-cast film was adopted, which used general organic solvent with relative low harmful effect at high temperature, inducing a spontaneous growth structure at the interface. **Figure 19(a)** and **(b)** shows AFM images of spin-cast film of diamide derivative and their composite with 1 wt% organo-modified MMT, respectively. From the comparison between these figures, it is confirmed that essential entanglement and wavy fibers have been performed the one-dimensional growth and changed to the linear shape in the spin-cast film by organo-MMT addition. As shown the results of XRD of **Figure 20** in the Supporting Information, a organo-MMT is also the developed layer-organization with 38 Å double-layered period [50]. Here, it will consider the interaction between organo-modified MMT and two-chain-type diamide derivative.

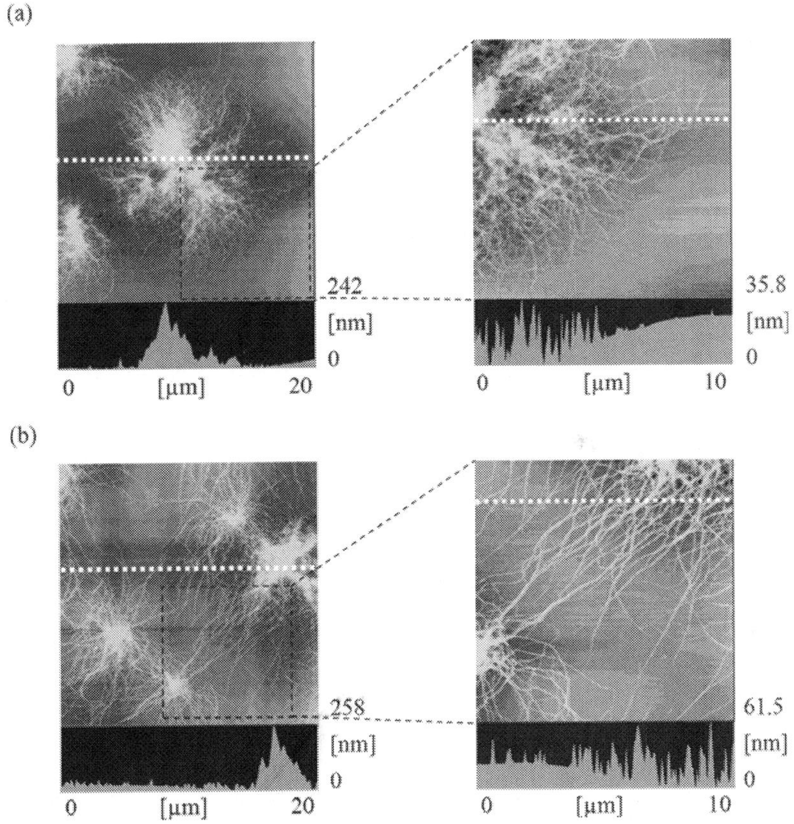

Figure 19. AFM images of spin-cast films of (a) neat diamide derivatives having two hydrocarbons and (b) their composite with 1 wt% organo-modified MMT.

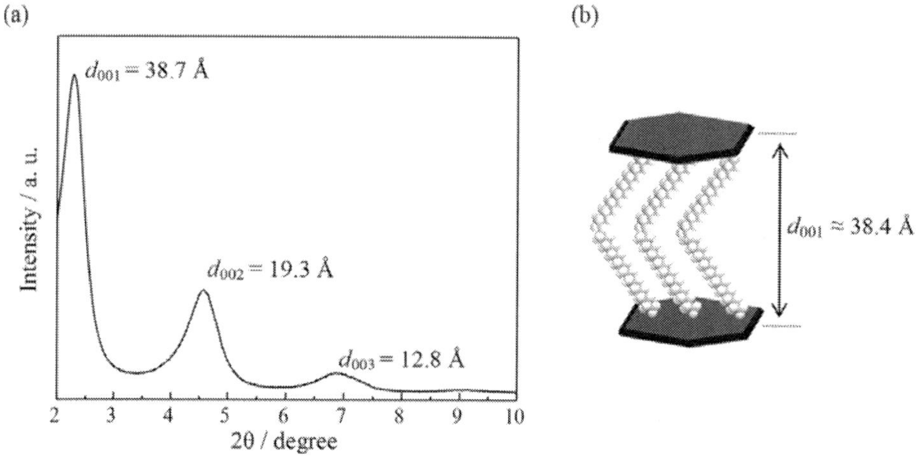

Figure 20. (a) Powder X-ray diffraction profile of organo-modified MMT used in this study [50]. (b) Schematic illustration of layer structure of organo-modified MMT in bulk.

Figure 21(a) indicates the π–A isotherms of the mixed monolayer of two-chain-type diamide derivative: organo-modified MMT = 3:1 and 1:1. Compared to behavior of an each original monolayers, since a change in the collapsed surface pressure occurs, it can be understood the existence of interaction between components. In AFM observation to this mixed monolayer transferred on solid substrate, the fibers of diamide derivative are in contact with the organo-MMT without phase separation, which can be understood to be a system of an ideal mixing (**Figure 21(b)**). Next, the results of XRD of spin-cast films of **Figure 19** are shown (**Figure 22(a)**). (Here, since the presence of the organo-MMT is a very small amount, they are not observed as the IR signal and this spectrum indicates very similar features to that in bulk as shown the IR result of **Figure 23** in the Supporting Information.) From the comparison between both profiles, it finds that the third-order reflection has appeared and the intensity of the second-order reflection slightly decreased by organo-MMT addition. Considering from the concept of odd–even effect of high-order reflection, it seems that "an expression of the higher-order reflection" and "the occurrence of odd–even tendency" indicate an enhancement of regularity (**Figure 22(b)**). As the supporting of this speculation, the intensity of out-of-plane XRD profile of diamide derivative–organo-clay = 1:1 mixed multilayers clearly is increased to the one of the film of single diamide derivative, which is based on the enhancement of regularity along the *c*-axis by addition of organo-clay (**Figure 22(c)**). A schematic diagram which summarizes the above discussion has shown in **Figure 24**. That is to say, it is found that the epitaxial growth along the (0 0 *l*) plane based on the similar long-spacing value of about 4 nm is induced from the affinity of diamide derivative and the long-alkyl chain on the montmorillonite surface.

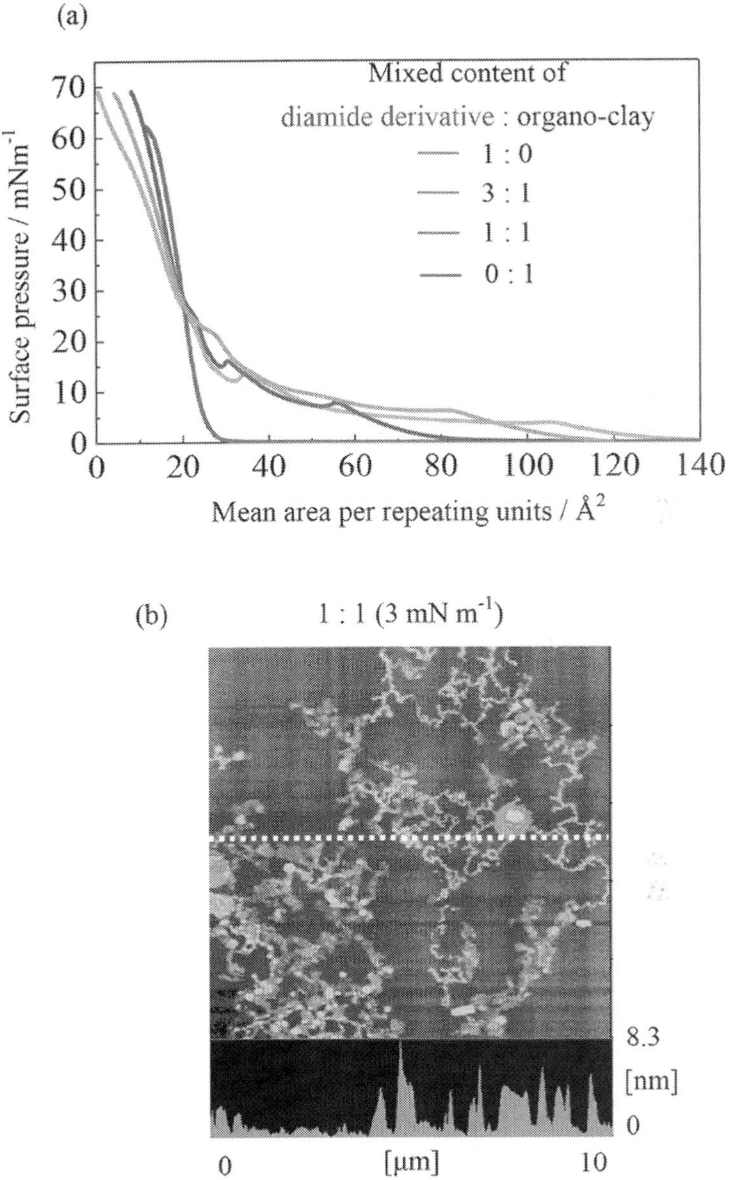

Figure 21. (a) Surface pressure–area isotherms of mixed monolayer on an aqueous buffer solution with Na^+ ion of diamide derivatives having two hydrocarbons:organo-MMT = 1:1 at 12°C. (b) AFM image of Z-type mixed monolayer on solid of diamide derivatives having two hydrocarbons–organo-MMT = 1:1 transferred at 3 mN m^{-1}.

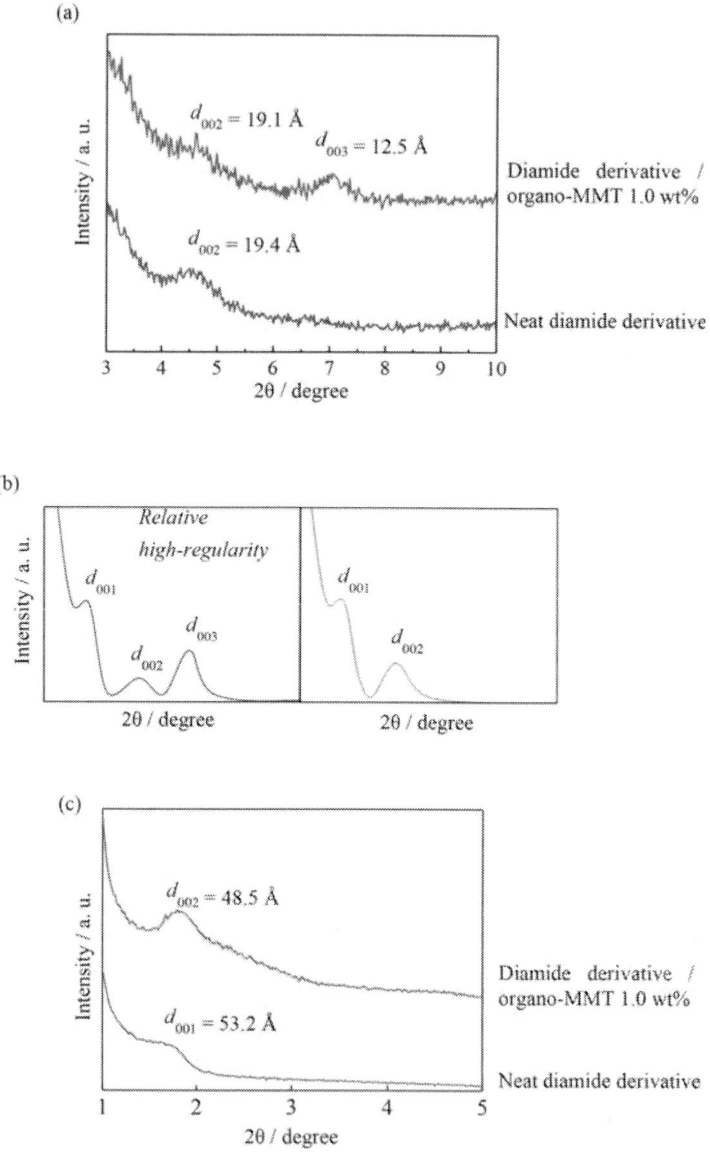

Figure 22. (a) XRD profiles of spin-cast films of neat diamide derivatives having two hydrocarbons and their composite with 1 wt% organo-modified MMT. (b) Simulation of XRD profiles of the layered materials influenced by an odd–even effect. (c) Out-of-plane XRD profiles of LB multilayers of neat diamide derivatives having two hydrocarbons and 1:1 mixed monolayer with organo-modified MMT.

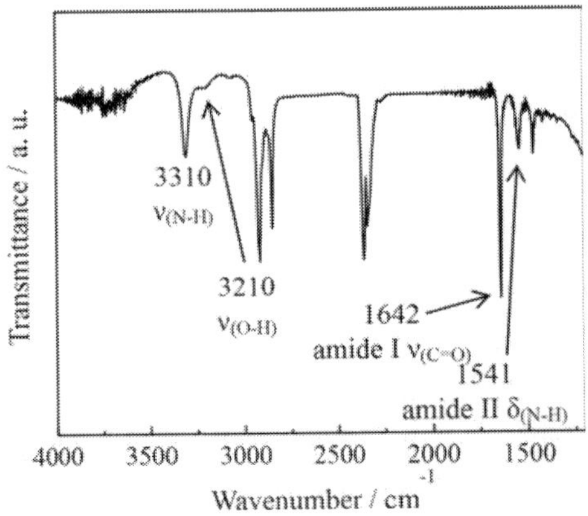

Figure 23. IR spectrum of mixed spin-cast film of two-chain-type diamide derivative with 1 wt% organo-modified MMT.

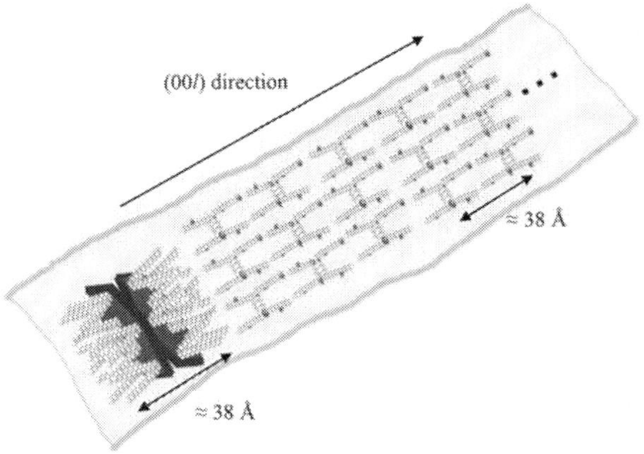

Figure 24. Schematic illustration of epitaxial growth of nanofiber morphology of two-chain-type diamide derivative from organo-modified MMT in spin-cast film.

2.4. Conclusions

In this chapter, monolayer behavior on the water surface, mesoscopic morphological formation, and molecular arrangement in LB multilayers of the diamide derivative having two hydrocarbons were investigated. For this purpose, π–A isotherms, in-plane and out-of-plane XRD, and AFM measurements were carried out. From the AFM analysis, the formation of nanofiber morphology of the thixotropic molecular organization was confirmed at different surface pressures. The in-plane and out-of-plane XRD measurements elucidated the formation of highly ordered layered structures and close-packing of molecular chains due to van der Waals interaction.

Nanofibers of the two-chain-type diamide derivative were obtained by applying the epitaxial growth method in spin-cast film. Under these conditions, the formation of linearly developed nanofibers was also observed. These mesoscopic extended nanofibers, whose internal structure consists of the packed long-alkyl chain, are expected to display cooperative thixotropic ability by fiber growth. Hence, these materials are likely candidates for future innovation in an additive, waste oil treatment agent, and paint industries.

Acknowledgements

The authors greatly appreciate Mr. Eiichi Sato, Kusumoto Chemicals Ltd., and Dr. Yoko Tatewaki, Tokyo University of Agriculture and Technology for their kind providing of samples. The authors also thank Mr. Kyohei Ohmura and Mr. Takahiro Kikkawa, Saitama University for his kind help of data analysis.

Author details

Manami Iizuka[1], Rie Yamato[2] and Atsuhiro Fujimori[1*]

*Address all correspondence to: fujimori@fms.saitama-u.ac.jp

1 Graduate School of Science and Engineering, Faculty of Engineering, Saitama University, Shimo-Okubo, Saitama, Japan

2 Department of Functional Materials Science, Faculty of Engineering, Saitama University, Shimo-Okubo, Saitama, Japan

References

[1] Blodgett K. Monomolecular Films of Fatty Acids on Glass. J. Am. Chem. Soc. 1934;56: 495–495. doi:10.1021/ja01317a513

[2] Kumaki J, Kawauchi T, Yashima E. Two-Dimensional Folded Chain Crystals of a Synthetic Polymer in a Langmuir–Blodgett Film. J. Am. Chem. Soc. 2005;127:5788–5789. doi:10.1021/ja050457e

[3] Mann B, Kuhn H. Tunneling through Fatty Acid Salt Monolayers. J. Appl. Phys. 1971;42:4398–4405. doi:10.1063/1.1659785

[4] Mann B, Kuhn H, Szentpály Lv. Tunnelling through Fatty Acid Monolayers and its Relevance to Photographic Sensitization. Chem. Phys. Lett. 1971;8:82–84. DOI: 10.1016/0009-2614(71)80582-1

[5] Onouchi H, Okoshi K, Kajitani T, Sakurai S, Nagai K, Kumaki J, Onitsuka K, Yashima E. Two- and Three-Dimensional Smectic Ordering of Single-Handed Helical Polymers. J. Am. Chem. Soc. 2008;130:229–236. doi:10.1021/ja074627u

[6] Kumaki J, Kajitani T, Nagai K, Okoshi K, Yashima E. Visualization of Polymer Chain Conformations in Amorphous Polyisocyanide Langmuir–Blodgett Films by Atomic Force Microscopy. J. Am. Chem. Soc. 2010;132:5604–5606. doi:10.1021/ja908426u

[7] Wu C, Li M, Chi S, Jun G. Laser Light Scattering Characterization of a Novel Polymer Nanofiber. Macromolecules 1998;31:7553–7554. doi:10.1021/ma9800663

[8] Huang MH, Mao S, Feick H, Yan H, Wu Y, Kind H, Weber E, Russo R, Yang P. Room-Temperature Ultraviolet Nanowire Nanolasers. Science 2001;292:1897–1899. doi: 10.1126/science.1060367

[9] Bumm LA, Arnold JJ, Cygan MT, Dunbar TD, Burgin TP, Jones L, Allara DL, Tour JM, Weiss PS. Are Single Molecular Wires Conducting? Science 1996;271:1705–1707. doi: 10.1126/science.271.5256.1705

[10] Fujimori A, Kaneko Y, Kikkawa T, Chiba S, Shibasaki Y. Fabrication and Structure of "Polymer Nanosphere Multilayered Organization". J. Colloid Interf. Sci. 2014;418:338–349. doi:10.1016/j.jcis.2013.12.038

[11] Park K, Sung Y, Han S, Yun Y, Hyeon T. Origin of the Enhanced Catalytic Activity of Carbon Nanocoil-Supported PtRu Alloy Electrocatalysts. J. Phys. Chem. B 2004;108:939–944. doi:10.1021/jp0368031

[12] Liu Y, Li H, Tu D, Ji Z, Wang C, Tang Q, Liu M, Hu W, Liu Y, Zhu D. Controlling the Growth of Single Crystalline Nanoribbons of Copper Tetracyanoquinodimethane for the Fabrication of Devices and Device Arrays. J. Am. Chem. Soc. 2006;128:12917–12922. doi:10.1021/ja0636183

[13] Johnston E, Bullock S, Uilk J, Gatenholm P, Wynne K. Networks from α,ω-Dihydroxypoly(dimethylsiloxane) and (Tridecafluoro-1,1,2,2-tetrahydrooctyl) triethoxysilane: Surface Microstructures and Surface Characterization. Macromolecules 1999;32:8173–8182. doi:10.1021/ma990628y

[14] Masuya R, Ninomiya N, Fujimori A, Nakahara H, Masuko T. Fine Structure and Phase Transition Behavior of Fluorinated Comb Copolymers. J. Polym. Sci. B, Polym. Phys. 2006;44:416–425. doi:10.1002/polb.20711

[15] Klok H, Langenwalter J, Lecommandoux S. Self-Assembly of Peptide-Based Diblock Oligomers. Macromolecules 2000;33:7819–7826. doi:10.1021/ma0009606

[16] Liu F, Prehm M, Zeng X, Tschierske C, Ungar G. Skeletal Cubic, Lamellar, and Ribbon Phases of Bundled Thermotropic Bolapolyphiles. J. Am. Chem. Soc. 2014;136:6846–6849. doi:10.1021/ja502410e

[17] Komiyama H, Sakai R, Hadano S, Asaoka S, Kamata K, Iyoda T, Komura M, Yamada T, Yoshida H. Enormously Wide Range Cylinder Phase of Liquid Crystalline PEO-*b*-PMA(Az) Block Copolymer. Macromolecules 2014;47:1777–1782. doi:10.1021/ma402356z

[18] Yabu H, Jia R, Matsuo Y, Ijiro K, Yamamoto S, Nishino F, Takaki T, Kuwahara M, Shimomura M. Preparation of Highly Oriented Nano-Pit Arrays by Thermal Shrinking of Honeycomb-Patterned Polymer Films. Adv. Mater. 2008;20:4200–4204. doi:10.1002/adma.200801170

[19] Tatewaki Y, Hatanaka T, Tsunashima R, Nakamura T, Kimura, M, Shirai H. Conductive Nanoscopic Fibrous Assemblies Containing Helical Tetrathiafulvalene Stacks. Chem. An Asian J. 2009;4:1474–1479. doi:10.1002/asia.200900044

[20] Tatewaki Y, Watanabe T, Watanabe K, Kikuchi K, Okada S. Synthesis and Nanostructures of Several Tetrathiafulvalene Derivatives Having the Side Chains Composed of Chiral and Hydrogen-Bonding Groups and their Charge-transfer Complexes. Dalton Trans. 2013;42:16121–16127. doi:10.1039/C3DT51464A

[21] Gaines Jr, G. L. Insoluble Monolayers at Liquid Gas Interfaces; Wiley: New York, 1966.

[22] Kuhn H, Möbius D, Bücher H. Spectroscopy of Monolayer Assemblies, In Physical Methods of Chemistry; Weissberger, A.; Rossiter, B. W., Eds.; Wiley: New York, 1972, Vol. 1, Part IIIB, p. 577–702.

[23] (a) Roberts GG. Langmuir–Blodgett Films; Plenum Press: London, 1990. ISBN: 978-1-4899-3716-2; (b) Petty MC. Langmuir–Blodgett Films; Cambridge Univ. Press: New York, 1996. ISBN: 9780521424509

[24] Ulman A. An Introduction to Ultrathin Organic Films: from Langmuir–Blodgett to Self-assembly; Academic Press: Boston, 1991. ISBN : 9780127082301

[25] Fukuda K, Shiozawa T. Conditions for Formation and Structural Characterization of X-type and Y-type Multilayers of Long-chain Esters. Thin Solid Films 1980;68:55–66. doi:10.1016/0040-6090(80)90136-4

[26] Fukuda K, Shiozawa T. Structural Control and Characterization in Mixed and Alternate Langmuir–Blodgett Films of Simple Long-chain Compounds. Thin Solid Films 1989;178:421–425. doi:10.1016/0040-6090(89)90332-5

[27] Fukuda K, Shibasaki Y, Nakahara H. Effects of Molecular Arrangement on Polymerization Reactions in Langmuir–Blodgett Films. Thin Solid Films 1983;99:87–94. doi:10.1016/0040-6090(83)90363-2

[28] Fukuda K, Shibasaki Y, Nakahara H. Molecular Arrangement and Polymerizability of Amino Acid Derivatives and Dienoic Acid in Langmuir–Blodgett Films. Thin Solid Films 1985;133:39–49. doi:10.1016/0040-6090(85)90423-7

[29] Fukuda K, Shibasaki Y, Nakahara H. Polymerizabilities of Amphiphilic Monomers with Controlled Arrangements in Langmuir–Blodgett Films. Thin Solid Films 1988;160:43–52. doi:10.1016/0040-6090(88)90045-4.

[30] Spooner SP, Whitten DG. Photoreactions in Monolayer Films and Langmuir–Blodgett Assemblies, In Photochemistry in Organized & Constrained Media; Ramamurthy, V., Ed.; VCH Pub.: New York, 1991. Chap 15. 691–738 p.

[31] Fujimori A, Sugita Y, Nakahara H, Ito E, Hara M, Matsuie N, Kanai K, Ouchi Y, Seki K. Change of Molecular Packing and Orientation from Monolayer to Multilayers of Hydrogenated and Fluorinated Carboxylates Studied by In-plane X-ray Diffraction together with NEXAFS Spectroscopy at CK-edge. Chem. Phys. Lett. 2004;387:345–348. doi:10.1016/j.cplett.2004.02.032

[32] Fujimori A, Araki T, Nakahara H, Ito E, Hara M, Ishii H, Ouchi Y, Seki K. In-plane X-ray Diffraction and Polarized NEXAFS Spectroscopic Studies on the Organized Molecular Films of Fluorinated Amphiphiles with Vinyl Esters and their Comb-polymers. Chem. Phys. Lett. 2001;349:6–12. doi:10.1016/S0009-2614(01)01196-4

[33] Tatewaki Y, Okada S, Itagaki R, Nakamura T, Fujimori A. Study of Molecular Arrangement of Organized Molecular Films of Charge-Transfer Complexes Containing 1, 3-Dithiole-2-Thione-4, 5-Dithiolate by In-plane and Out-of-plane X-ray Diffractions. J. Colloid Interf. Sci. 2010;343:281–290. doi:10.1016/j.jcis.2009.09.031

[34] Jung JH, Ono Y, Shinkai S. Sol-Gel Polycondensation in a Cyclohexane-Based Organogel System in Helical Silica: Creation of both Right- and Left-Handed Silica Structures by Helical Organogel Fibers. Chem. Eur. J. 2000;6:4552–4557. doi: 10.1002/1521-3765(20001215)6:24<4552::AID-CHEM4552>3.0.CO;2-5;

[35] Jung JH, Ono Y, Shinkai S. Sol-Gel Polycondensation of Tetraethoxysilane in a Cholesterol-Based Organogel Synstem Results in Chiral Spiral Silica. Angew. Chem. Int. Ed. 2000;39:1862–1865. doi:10.1002/(SICI)1521-3773(20000515)39:10<1862::AID-ANIE1862>3.0.CO;2-3;

[36] Jung JH, Kobayashi H, Mitsutoshi M, Shimizu T, Shinkai S. Helical Ribbon Aggregate Composed of a Crown-Appended Cholesterol Derivative Which Acts as an Amphiphilic Gelator of Organic Solvents and as a Template for Chiral Silica Transcription. J. Am. Chem. Soc. 2001;123:8785–8789. doi:10.1021/ja010508h.

[37] Jha SK, Cheon K, Green MM, Selinger JV. Chiral Optical Properties of a Helical Polymer Synthesized from Nearly Racemic Chiral Monomers Highly Diluted with Achiral Monomers. J. Am. Chem. Soc. 1999;121:1665–1673. doi:10.1021/ja983202s

[38] Green MM, Khatri C, Peterson NC. A Macromolecular Conformational Change Driven by a Minute Chiral Solvation Energy. J. Am. Chem. Soc. 1993;115:4941–4942. doi: 10.1021/ja00064a086

[39] Konuma T, Akutagawa T, Yumoto T, Nakamura T, Kawamata J, Inoue K, Nakamura T, Tachibana H, Matsumoto M, Ikegami K, Horiuchi H, Yamochi H, Saito G. Charge-transfer Interactions and Non-linear Optical Properties of Tetrathiafulvalene-based Langmuir–Blodgett Films. Thin Solid Films 1998;327–329:348–352. DOI:10.1016/S0040-6090(98)00667-1

[40] Horiuchi S, Yamochi H, Saito G, Sakaguchi K, Kusunoki M. Nature and Origin of Stable Metallic State in Organic Charge-Transfer Complexes of Bis(ethylenedioxy)tetrathia-fulvalene. J. Am. Chem. Soc. 1996;118:8604–8622. doi:10.1021/ja960393v

[41] Torrance JB, Scot BA, Welber B, Kaufman FB, Seiden PE. Optical Properties of the Radical Cation Tetrathiafulvalenium (TTF^{+}) in its Mixed-valence and Monovalence Halide Salts. Phys. Rev. B: Condens Matter 1979;19:730–741. doi:10.1103/PhysRevB.19.730

[42] Bradshaw D, Claridge JB, Cussen EJ, Prior TJ, Rosseinsky MJ. Design, chirality, and flexibility in nanoporous molecule-based materials. Accounts Chem. Res. 2005;38:273–282. doi:10.1021/ar0401606

[43] Barnes HA. Thixotropy—A review. J. Non-Newtonian Fluid Mech. 1997;70:1–33. doi: 10.1016/S0377-0257(97)00004-9

[44] Karimi-Maleh H, Tahernejad-Javazm F, Ensafi AA, Moradi R, Reza, Mallakpour S, Beitollahi H. A high sensitive biosensor based on FePt/CNTs nanocomposite/N-(4-hydroxyphenyl)-3,5-dinitrobenzamide modified carbon paste electrode for simultaneous determination of glutathione and piroxicam. Biosens. Bioelectron. 2014;60:1–7. doi: 10.1016/j.bios.2014.03.055

[45] Karimi-Maleh H, Biparva P, Hatami M. A novel carbon paste electrode based on NiO/CNTs nanocomposite and (9,10-dihydro-9,10-ethanoanthracene-11,12-dicarboximido)-4-ethylbenzene-1,2-diol as a mediator for simultaneous determination of cysteamine, nicotinamide adenine dinucleotide and folic acid. Biosens. Bioelectron. 2013;48:270–275. doi:10.1016/j.bios.2013.04.029

[46] Karimi-Maleh H, Tahernejad-Javazmi F, Atar N, Lutfi M, Gupta VK, Ensafi AA. A novel DNA biosensor based on a pencil graphite electrode modified with polypyrrole/functionlized multiwalled carbon nanotubes for determination of 6-mercaptopurine anticancer drug. 2015;54:3634–3639. doi:10.1021/ie504438z

[47] Moradi R, Sebt SA, Karimi-Maleh H, Sadeghi Roya, Karimi F, Bahari A, Arabi H. Synthesis and application of FePt/CNTs nanocomposite as a sensor and novel amide ligand as a mediator for simultaneous determination of glutathione, nicotinamide adenine dinucleotide and tryptophan. Phys. Chem. Chem. Phys. 2013;15:5888-5897 DOI: 10.1039/C3CP00033H

[48] Fujimori A, Yamato R, Kikkawa T, Tatewaki Y. Morphological Transition of a Conductive Molecular Organization with Non-covalent from Nanonetwork to Nanofiber. J. Colloid Interf. Sci. 2015;448:180–188. doi:10.1016/j.jcis.2015.02.024

[49] Fujimori A, Arai S, Kusaka J, Kubota M, Kurosaka K. Formation and Structure of Langmuir–Blodgett Films of Organo-modified Alminosilicate with High Surface Coverage. J. Colloid Interf. Sci. 2013;392:256–265. doi:10.1016/j.jcis.2012.08.075

[50] Fujimori A, Kusaka J, Nomura R. Formation and Structure of Organized Molecular Films for Organo-modified Montmorillonite and Mixed Monolayer Behavior with Poly(L-lactide). Polym. Eng. Sci. 2011;51:1099–1107. doi:10.1002/pen.21912